D0720787

RESPONSIBLE RESEARCH

WITH BIOLOGICAL SELECT AGENTS AND TOXINS

Committee on Laboratory Security and Personnel
Reliability Assurance Systems for Laboratories Conducting
Research on Biological Select Agents and Toxins

Board on Life Sciences

Division on Earth and Life Studies

NATIONAL RESEARCH COUNCIL
OF THE NATIONAL ACADEMIES

THE NATIONAL ACADEMIES PRESS
Washington, D.C.
www.nap.edu

THE NATIONAL ACADEMIES PRESS   500 Fifth Street, NW   Washington, DC 20001

NOTICE: The project that is the subject of this report was approved by the Governing Board of the National Research Council, whose members are drawn from the councils of the National Academy of Sciences, the National Academy of Engineering, and the Institute of Medicine. The members of the committee responsible for the report were chosen for their special competences and with regard for appropriate balance.

This study was supported by Contract No. N01-OD-4-2139 (Task Order #218) between the National Academy of Sciences and the National Institutes of Health. The content of this publication does not necessarily reflect the views or policies of the Department of Health and Human Services, nor does mention of trade names, commercial products, or organizations imply endorsement by the U.S. Government.

International Standard Book Number-13: 978-0-309-14535-0 (Book)
International Standard Book Number-10: 0-309-14535-X (Book)
International Standard Book Number-13: 978-0-309-14536-7 (PDF)
International Standard Book Number-10: 0-309-14536-8 (PDF)
Library of Congress Control Number: 2009940166

Additional copies of this report are available from the National Academies Press, 500 Fifth Street, NW, Lockbox 285, Washington, DC 20055; (800) 624-6242 or (202) 334-3313 (in the Washington metropolitan area); Internet, http://www.nap.edu.

Copyright 2009 by the National Academies. All rights reserved.

Printed in the United States of America

QR
64.7
.N387
2009

THE NATIONAL ACADEMIES
Advisers to the Nation on Science, Engineering, and Medicine

The **National Academy of Sciences** is a private, nonprofit, self-perpetuating society of distinguished scholars engaged in scientific and engineering research, dedicated to the furtherance of science and technology and to their use for the general welfare. Upon the authority of the charter granted to it by the Congress in 1863, the Academy has a mandate that requires it to advise the federal government on scientific and technical matters. Dr. Ralph J. Cicerone is president of the National Academy of Sciences.

The **National Academy of Engineering** was established in 1964, under the charter of the National Academy of Sciences, as a parallel organization of outstanding engineers. It is autonomous in its administration and in the selection of its members, sharing with the National Academy of Sciences the responsibility for advising the federal government. The National Academy of Engineering also sponsors engineering programs aimed at meeting national needs, encourages education and research, and recognizes the superior achievements of engineers. Dr. Charles M. Vest is president of the National Academy of Engineering.

The **Institute of Medicine** was established in 1970 by the National Academy of Sciences to secure the services of eminent members of appropriate professions in the examination of policy matters pertaining to the health of the public. The Institute acts under the responsibility given to the National Academy of Sciences by its congressional charter to be an adviser to the federal government and, upon its own initiative, to identify issues of medical care, research, and education. Dr. Harvey V. Fineberg is president of the Institute of Medicine.

The **National Research Council** was organized by the National Academy of Sciences in 1916 to associate the broad community of science and technology with the Academy's purposes of furthering knowledge and advising the federal government. Functioning in accordance with general policies determined by the Academy, the Council has become the principal operating agency of both the National Academy of Sciences and the National Academy of Engineering in providing services to the government, the public, and the scientific and engineering communities. The Council is administered jointly by both Academies and the Institute of Medicine. Dr. Ralph J. Cicerone and Dr. Charles M. Vest are chair and vice chair, respectively, of the National Research Council.

www.national-academies.org

COMMITTEE ON LABORATORY SECURITY AND PERSONNEL RELIABILITY ASSURANCE SYSTEMS FOR LABORATORIES CONDUCTING RESEARCH ON BIOLOGICAL SELECT AGENTS AND TOXINS

RITA R. COLWELL (*Chair*), Distinguished University Professor, University of Maryland, College Park, MD, and Johns Hopkins University Bloomberg School of Public Health, Baltimore, MD; and President and Chief Executive Officer, CosmosID, Inc., Bethesda, MD

RONALD M. ATLAS, Professor of Biology and Public Health and Co-Director, Center for Health Preparedness, University of Louisville, Louisville, KY

JOHN D. CLEMENTS, Professor and Chair, Department of Microbiology and Immunology, and Director, Tulane Center for Infectious Diseases, Tulane University, New Orleans, LA

JOSEPH A. DiZINNO, Technical Director, Homeland Security and Law Enforcement, BAE Systems, Washington, DC

ADOLFO GARCÍA-SASTRE, Professor of Microbiology, Fischberg Chair and Professor of Medicine, and Co-Director, Global Health and Emerging Pathogens Institute, Mount Sinai School of Medicine, New York, NY

MICHAEL G. GELLES, Senior Manager, Deloitte Consulting LLP, Washington, DC

ROBERT J. HAWLEY, Senior Advisor for Science, Midwest Research Institute, Frederick, MD

SALLY KATZEN, Executive Managing Director, The Podesta Group, Washington, DC

PAUL LANGEVIN, Director of Laboratory Design, Merrick and Company, and President, Merrick Canada ULC, Kanata, Ontario, Canada

TODD R. LaPORTE, Professor Emeritus of Political Science, University of California, Berkeley, CA

STEPHEN S. MORSE, Professor of Clinical Epidemiology and Founding Director, Center for Public Health Preparedness, Columbia University Mailman School of Public Health, New York, NY

KATHRYN NEWCOMER, Professor and Director, Trachtenberg School of Public Policy and Public Administration, and Co-Director, Midge Smith Center for Evaluation Effectiveness, George Washington University, Washington, DC

ELIZABETH RINDSKOPF PARKER, Dean, McGeorge School of Law, University of the Pacific, Sacramento, CA

PAUL R. SACKETT, Beverly and Richard Fink Distinguished Professor of Psychology and Liberal Arts, University of Minnesota, Minneapolis, MN

Staff

ADAM P. FAGEN, Study Director and Senior Program Officer
JO L. HUSBANDS, Scholar/Senior Project Director
RITA GUENTHER, Senior Program Associate
CARL-GUSTAV ANDERSON, Senior Program Assistant

BOARD ON LIFE SCIENCES

KEITH YAMAMOTO (*Chair*), University of California, San Francisco, CA
ANN M. ARVIN, Stanford University, Stanford, CA
BONNIE L. BASSLER, Princeton University, Princeton, NJ
VICKI L. CHANDLER, Gordon and Betty Moore Foundation, Palo Alto, CA
SEAN EDDY, Janelia Farm Research Campus, Howard Hughes Medical Institute, Ashburn, VA
MARK D. FITZSIMMONS, John D. and Catherine T. MacArthur Foundation, Chicago, IL
DAVID R. FRANZ, Midwest Research Institute, Frederick, MD
LOUIS J. GROSS, University of Tennessee, Knoxville, TN
JO HANDELSMAN, University of Wisconsin, Madison, WI
CATO T. LAURENCIN, University of Connecticut Health Center, Farmington, CT
JONATHAN D. MORENO, University of Pennsylvania, Philadelphia, PA
ROBERT M. NEREM, Georgia Institute of Technology, Atlanta, GA
CAMILLE PARMESAN, University of Texas, Austin, TX
MURIEL E. POSTON, Skidmore College, Saratoga Springs, NY
ALISON G. POWER, Cornell University, Ithaca, NY
BRUCE W. STILLMAN, Cold Spring Harbor Laboratory, Cold Spring Harbor, NY
CYNTHIA WOLBERGER, Johns Hopkins University School of Medicine, Baltimore, MD
MARY WOOLLEY, Research!America, Alexandria, VA

Staff

FRANCES E. SHARPLES, Director
JO L. HUSBANDS, Scholar/Senior Project Director
ADAM P. FAGEN, Senior Program Officer
ANN H. REID, Senior Program Officer
MARILEE K. SHELTON-DAVENPORT, Senior Program Officer
INDIA HOOK-BARNARD, Program Officer
ANNA FARRAR, Financial Associate
CARL-GUSTAV ANDERSON, Senior Program Assistant
AMANDA P. CLINE, Senior Program Assistant
AMANDA MAZZAWI, Program Assistant

Preface

As a scientist who has worked for more than 40 years to find cures for infectious disease, I find the idea that terrorists would use biological agents as a weapon to be anathema. It violates the fundamental values of the life sciences that I and my colleagues hold dear: that science is a vital tool for improving life and the health of our planet and enhancing our understanding of the natural world.

My own work has focused on cholera, a disease responsible for the death of thousands of people around the world every year. During the past 40 years, research carried out through international collaboration of scientists has saved many thousands of lives.

At the same time, we are firm in the belief that this research should be conducted safely and responsibly. The incidence of either laboratory workers or members of the public being infected is vanishingly small, whether from laboratory accidents or intentional action. Through the years, safety and security practices and procedures have been developed that have successfully prevented accidental or intentional misuse of biological materials.

While research with select agents and toxins introduces another level of potential risk, the same sense of responsibility applies. Scientists have not only demonstrated concern about these issues, but also recognize that they have the most at stake should an incident occur. They are best able to identify potential risk, whether from a laboratory door left unsecured or the unusual behavior of a laboratory worker. It is for these reasons that this report focuses on promoting a culture of responsibility, enabling and empowering scientists to be vigilant stewards of their science.

Research with select agents and toxins is both necessary and important. Our nation's health and security depend upon our understanding of these potentially dangerous pathogens and their mechanisms of virulence. Our fundamental understanding of life and life processes benefits from study of these agents. Nevertheless, there is the possibility that we can be overzealous, implementing procedures only thought to enhance security. While many current policies and practices are effective, some actions suggested to enhance security are not likely to make select agent research more secure, just more difficult to conduct; this may yield the opposite result: that overall security will be diminished, not strengthened.

The authoring committee for this report represents a broad cross-section of stakeholders, including select agent researchers, experts in psychology, professionals in biosafety and facility design, and individuals with extensive experience in the issues of science and security. The report represents a consensus of the committee and our best judgment on the most effective ways to both promote security and foster scientific knowledge and a rapid biological response in the event of an emergency.

With such a challenging task, the committee was given only 3½ months to complete a full report. As such, the committee had to make choices about which issues to address, concentrating on those it felt to be most important, most critical, and most effective for enhancing security and enabling research. Thanks to the dedication of both the committee and staff, analysis of the issues included in the report can be considered no less thorough and documented than if we had been given the luxury of time. The study was conducted at the request of the leadership of the National Interagency Biodefense Campus and the White House Homeland Security Council staff through a contract with the National Institute of Allergy and Infectious Diseases.

On behalf of the entire committee, I wish to extend our sincere gratitude to the excellent staff at the National Academies. This report represents a full year's worth of work conducted in less than four months. It is because of the dedication and extraordinary efforts of study director Adam Fagen, Jo Husbands, Rita Guenther, and Carl-Gustav Anderson that we were able to complete this ambitious task is so short a time. The staff most impressively captured the conclusions of the committee's discussions and ensured access to the information and expertise we needed. The committee was able to identify the most important issues and reach consensus with relative ease because of the superb work of the staff. The tasks were facilitated by a knowledgeable, dedicated, and insightful committee, and I thank my fellow committee members for their commitment that made the study process an enjoyable and rewarding opportunity.

In closing, "every researcher, whether in academia, in government research facilities, or in industry, needs to be aware of the potential unintended consequences of their own and their colleagues' research. In 1975, scientists agreed to the 'Asilomar moratorium,' which gave guidance to researchers in the emerging field of recombinant DNA research. Today, researchers in the biological sciences again need to take responsibility for helping to prevent the potential misuses of their work, while being careful to preserve the vitality of their disciplines as required to contribute to human welfare."[1] The committee sincerely hopes that its work will contribute usefully to ongoing discussion of the Select Agent Program and, especially, to the safety and security of select agent research.

Rita R. Colwell, Chair

---

[1]Bruce Alberts and Robert M. May. 2002. Scientist Support for Biological Weapons Controls. *Science* 298(November 8): 1135.

Acknowledgments

This report has been reviewed in draft form by individuals chosen for their diverse perspectives and technical expertise, in accordance with procedures approved by the National Academies' Report Review Committee. The purpose of this independent review is to provide candid and critical comments that will assist the institution in making its published report as sound as possible and to ensure that the report meets institutional standards for objectivity, evidence, and responsiveness to the study charge. The review comments and draft manuscript remain confidential to protect the integrity of the process.

We wish to thank the following individuals for their review of this report:

Burt S. Barnow, *Johns Hopkins University*
W. Seth Carus, *National Defense University*
Wayne F. Cascio, *University of Colorado Denver*
Elizabeth Casman, *Carnegie Mellon University*
R. John Collier, *Harvard Medical School*
Nancy D. Connell, *University of Medicine and Dentistry of New Jersey*
Penny H. Holeman, *Lovelace Respiratory Research Institute*
Joseph Kanabrocki, *University of Chicago*
Joseph Krofcheck, *Independent Consultant*
Thomas G. Ksiazek, *University of Texas Medical Branch at Galveston*
Admiral Mike McConnell, *Booz Allen Hamilton*
Denise A. Pettit, *Virginia Division of Consolidated Laboratory Services*
William H. Press, *University of Texas at Austin*
David A. Relman, *Stanford University*
John F. Sopko, *Akin Gump Strauss Hauer & Feld LLP*
Tilahun D. Yilma, *University of California, Davis*

Although the reviewers listed above have provided many constructive comments and suggestions, they were not asked to endorse the conclusions or recommendations, nor did they see the final draft of the report before its release. The review of this report was overseen by **W. Emmett Barkley**, *Proven Practices LLC*, and **David R. Challoner**, *University of Florida (emeritus)*. Appointed by the National Academies, they were responsible for making certain that an independent examination of this report was carried out in accordance with institutional procedures and that all review comments were carefully considered. Responsibility for the final content of this report rests entirely with the authoring committee and the institution.

The committee is grateful for those who provided expertise and assistance throughout the study process. This includes those experts who spoke to the committee at one of its meetings: Jeffrey Adamovicz, LouAnn Burnett, Sheldon Cohen, M. Colleen Crowley, Diane Damos, Robert Fein, Kelley Krokos, Bruce Landry, H. Clifford Lane, J. William Leonard, Carol Linden, Richard Meserve, Dennis Metzger, Kevin Murphy, Ben Petro, Mary Rowe, Bryan Vossekuil, Robbin Weyant, and Linda Wilcox. Meeting agendas and speaker affiliations are listed in Appendix B.

The committee is also thankful to those who helped organize or participate in one of the committee's site visits at the New England Regional Center of Excellence for Biodefense and Emerging Infectious Diseases Research at Harvard Medical School (Christine Anderson, Gerald Beltz, Mary Corrigan, Robert Dickson, Sara Heninger, Andrew Onderdonk, and Jeff Seo); the MIT Nuclear Reactor Laboratory (John Bernard, David Carpenter, Patricia Drooff, Edward Lau, William McCarthy, Thomas Newton, Jr., and Kathleen O'Connell) and Environmental, Health and Safety Office (Claudia Mickelson) at the Massachusetts Institute of Technology; George Mason University's National Center for Biodefense and Infectious Diseases (Saira Ahmad, Lilian Amer, Charles Bailey, John Blacksten, Calvin Carpenter, Jessica Chertow, Myung Chung, Meghan Durham-Colleran, Suhua Han, Jessica Kidd, Nathan Manes, Beth McKenney, Marjorie Musick, Tony Pierson, Kathleen Powell, Meena Rajan, Ian Reynolds, Diann Stedman, Anne Taylor, Patty Theimer, Monique van Hoek, Anne Verhoeven, Paul Wieber, James Willett, and Ron Witt); and the U.S. Department of Agriculture's National Plant Germplasm and Biotechnology Laboratory (Wayne Claus, Renee DeVries, Joseph Kozlovac, and Laurene Levy). A complete list of site visit participants and affiliations is available in Appendix B.

Thanks also to those who provided or facilitated access to additional information and input to the committee including Lida Anestidou, Dennis Ausiello, Charles Bailey, Kavita Berger, Steve Brooks, Matthew Burch, M. Colleen Crowley, David Tutrong Diec, Peter Emanuel, Deborah Glickstein, Gigi Kwik Gronvall, Bauke Houtman, James LeDuc, Carol Linden, Jean Patterson, Ben Petro, Paul Stern, Eric Utt, Raymond Webber, Robbin Weyant, and Carrie Wolinetz.

A factual review of Chapter 2 was conducted by the Department of Health and Human Services (Laura Kwinn and Carol Linden on behalf of the inter-agency working group on these issues), the Centers for Disease Control and Prevention (Robbin Weyant, Director of the Division on Select Agents and Toxins), the U.S. Department of Agriculture (coordinated by Julia Kiehlbauch in the Animal and Plant Health Inspection Service), and the Federal Bureau of Investigation (Edward You, Supervisory Special Agent on the Bioterrorism Team, and the staff of the Criminal Justice Information Service).

Contents

Executive Summary

Scientists have been conducting research with the organisms classified as biological select agents and toxins (BSAT) for several hundred years in order to understand the biology of these potentially dangerous pathogens and to develop countermeasures that will diminish the threat they pose. Because of legitimate concerns that BSAT materials might be used in deliberate criminal or terrorist acts, the federal government has instituted policies and procedures governing the security of BSAT laboratories and the reliability of personnel who work with BSAT materials. The committee was asked to consider the appropriate framework for laboratory security and personnel reliability measures that will optimize benefits, minimize risk, and facilitate the productivity of research.[1]

The committee identified six principles that should guide consideration of BSAT research; these principles also provide the lens through which the committee offers its conclusions and recommendations:

1. Research on biological select agents and toxins is essential to the national interest.
2. Research with biological select agents and toxins introduces potential security and safety concerns.
3. The Select Agent Program should focus on those biological agents and toxins that might be used as biothreat agents.
4. Policies and practices for work with biological select agents and toxins should promote both science and security.
5. Not all laboratories and not all agents are the same.
6. Misuse of biological materials is taboo in every scientific community.

---

[1] See Box 1-2 for the full statement of task.

Consideration of these principles led the committee to nine recommendations that it believes are essential for keeping BSAT research secure from both internal and external threats.

RECOMMENDATIONS

Recommendation 1 assigns responsibility for fostering a culture of trust and responsibility to a partnership of laboratory leaders and the Select Agent Program:

RECOMMENDATION 1: Laboratory leadership and the Select Agent Program should encourage and support the implementation of programs and practices aimed at fostering a culture of trust and responsibility within BSAT entities. These programs and practices should be designed to minimize potential security and safety risks by identifying and responding to potential personnel issues. These programs should have a number of common elements, tailored to reflect the diversity of facilities conducting BSAT research:

- **Consideration should be given to including discussion of personnel monitoring during (1) the initial training required for all personnel prior to gaining access to BSAT materials and annual refresher updates and (2) safety inspections to obtain a more complete assessment of the laboratory's ability to provide a safe and secure research environment.**
- **More broadly, personnel with access to select agents and toxins should receive training in scientific ethics and dual-use research. Training should be designed to foster community responsibility and raise awareness of all personnel of available institutional support and medical resources.**
- **Federal agencies overseeing and sponsoring BSAT research and professional societies should provide educational and training resources to accomplish these goals.**

Recommendation 2 engages the research community in oversight of the Select Agent Program through formation of an advisory committee:

RECOMMENDATION 2: To provide continued engagement of stakeholders in oversight of the Select Agent Program, a Biological Select Agents and Toxins Advisory Committee (BSATAC) should be established. The members, who should be drawn from academic/research institutions and the private sector, should include microbiologists and other infec-

tious disease researchers (including select agent researchers), directors of BSAT laboratories, and those with experience in biosecurity, animal care and use, compliance, biosafety, and operations. Representatives from the federal agencies with a responsibility for funding, conducting, or overseeing select agent research would serve in an ex officio capacity. Among the responsibilities of this advisory committee should be the following:

- Promulgate guidance on the implementation of the Select Agent Program;
- Facilitate exchange of information across institutions and sectors;
- Promote sharing of successful practices across institutions and sectors;
- Provide oversight for evaluation of the Select Agent Program;
- Provide advice on composition/stratification of the list of select agents and toxins;
- Convene regular meetings of key constituency groups; and
- Promote harmonization of regulatory policies and practices.

Two recommendations address the composition of the list of select agents and toxins and the implications that the nature of the agents has for accountability:

RECOMMENDATION 3: The list of select agents and toxins should be stratified in risk groups according to the potential use of the material as a biothreat agent, with regulatory requirements and procedures calibrated against such stratification. Importantly, mechanisms for timely inclusion or removal of an agent or toxin from the list are necessary and should be developed.

RECOMMENDATION 4: Because biological agents have an ability to replicate, accountability is best achieved by controlling access to archived stocks and working materials. Requirements for counting the number of vials or other such measures of the quantity of biological select agents (other than when an agent is transported from one laboratory site to another) should not be employed because they are both unreliable and counter-productive, yielding a false sense of security. A registered entity should record the identity of all biological select agents and toxins within that entity, where such materials are stored, who has access and when that access is available, and the intended use(s) of the materials.

There have been extensive discussions about the appropriateness of the current Security Risk Assessment process for screening personnel before they are permitted to work with BSAT; the committee concluded that this process is adequate for screening, but there should be an opportunity to consider mitigating factors as part of an appeal process:

> **RECOMMENDATION 5: The current Security Risk Assessment screening process should be maintained, but the appeal process should be expanded beyond the simple check for factual errors to include an opportunity to consider the circumstances surrounding otherwise disqualifying factors.**

Because of confusion within the community about how physical security requirements should be implemented, the committee calls upon the Select Agent Program to provide a minimum set of requirements that would apply across agencies:

> **RECOMMENDATION 6: The Select Agent Program should define minimum cross-agency physical security requirements, which recognize that facilities have unique risk-based security needs and associated design components, to assist facilities in meeting their regulatory obligations.**

The committee recognizes the importance of data to inform the operation of the Select Agent Program and recommends ongoing independent evaluation of the program:

> **RECOMMENDATION 7: Independent evaluation of the Select Agent Program should be undertaken to assess the relative benefits for achieving security, to consider the consequences of the program (intended and unintended) on the research enterprise, and to provide useful data about the Select Agent Program. Such evaluation, which may be coordinated through the BSAT Advisory Committee, should be provided with dedicated funding.**

Recognizing the critical role that laboratory inspections play in maintaining the efficient and effective operations of select agent research, the committee calls for appropriate knowledge, experience, and training among inspectors:

> **RECOMMENDATION 8: Inspectors of select agent laboratories should have scientific and laboratory knowledge and experience, as well as appropriate training in conducting inspections specific to BSAT research. Inspector training and practice should be harmonized across federal, state, local, and other agencies.**

Finally, the committee concluded that security and compliance costs have been a challenge for the BSAT research community and calls upon federal funding agencies to provide sustained support for these facility costs:

RECOMMENDATION 9: Because of considerable security and compliance costs associated with research on biological select agents and toxins, federal agencies funding BSAT research should establish a separate category of funding to ensure sustained support for facilities where such research is conducted.

Summary

INTRODUCTION

More than 300 years ago, Antonie van Leeuwenhoek constructed a primitive microscope and made the first detailed descriptions of microorganisms. More than 200 years ago, Edward Jenner carried out the first experimental vaccination, using cow pox virus to build immunity in humans against the deadly smallpox virus. More than 100 years ago, Robert Koch isolated the *Bacillus anthracis* bacterium and postulated a causal relationship between specific microorganisms and disease.

From these early discoveries, scientists have built more than a century of research on microorganisms and infectious disease, including research on some of the most dangerous pathogens. Enormous advances have resulted in the development of vaccines and other treatments that have greatly diminished the risks posed by infectious disease agents. It is not an exaggeration to attribute increased lifespan and better human health to the research of legions of microbiologists and other biomedical researchers on the biology of bacteria and viruses and the toxins they produce. At the same time, these researchers have maintained safety and responsibility in the laboratory. Notwithstanding the enormous volume of infectious disease research that has been accomplished, there have been few incidents of pathogenic organisms being released into the environment by accident, negligence, or deliberate action. Moreover, scientific research is safer than it has ever been because of increasing concern for safety and security and implementation of protective measures that minimize risk.

Among the large group of pathogenic materials is a smaller set of organisms and chemicals that pose not only a severe threat to the health of humans, plants, and animals, but also have the potential to be used deliberately to cause disease, prompt fear, or destroy agricultural or animal products. More than 80 of these most dangerous bacteria, viruses, toxins, and fungi have been officially listed by the U.S. government as biological select agents and toxins (BSAT) and are subject to special security requirements.

Whether deliberately deployed as a biological weapon or the result of a natural outbreak, the potential for mass human casualty or potentially catastrophic impact on plants or animals as a direct or indirect result of select agents is omnipresent. This report focuses on how to secure access to these dangerous pathogens to diminish their potential for use by terrorists as a biothreat agent. Discussion includes consideration of the physical security of facilities that work with these materials and steps to ensure that personnel with access to select agents and toxins can be trusted.

The Current Select Agent Program

Since the list of select agents and toxins was first introduced in 1997, the U.S. government has created a formal regulatory structure to oversee BSAT research and to decide who could possess microorganisms and toxins that could be used as weapons and how facilities that did possess them would be protected. The scope of the program is determined by a formal list of select agents and toxins; the Department of Health and Human Services' Centers for Disease Control and Prevention (CDC) maintains the list for human pathogens, while the U.S. Department of Agriculture's Animal and Plant Health Inspection Service (APHIS) maintains the list for plant and animal pathogens.

As of September 2009, 388 entities were registered and 13,609 individuals—administrators, research scientists, students and postdoctoral researchers, technical staff, and maintenance and animal care workers—were cleared to have access to BSAT materials.

Origin and Charge to the Committee

Concerns about whether the regulations in place for BSAT research in U.S. laboratories were adequate to address the risks of theft, misuse, or diversion of materials grew after the Federal Bureau of Investigation (FBI) announced in August 2008 that it had concluded that a researcher at the U.S. Army Medical Research Institute of Infectious Diseases was the perpetrator of the anthrax letter attacks in October 2001. There were also other concerns about whether the growth in the number of high containment laboratories as part of expanded funding for biodefense research after 2001 was increasing the risks of laboratory accidents as well as providing more targets for those who could pose security threats from either outside or inside the facilities.

An interagency process was initiated to consider the efficiency and effectiveness of all laws, regulations, guidance, and practices related to physical, facility, and personnel security and assurance for BSAT research. As part of that process, a government Working Group, created by an Executive Order (EO) issued by President George W. Bush, delivered its assessment to the President in July 2009. The Homeland Security Council staff requested additional input from the National Science Advisory Board for Biosecurity (NSABB) and the National Research Council (NRC).[1] This NRC report considers the efficacy of regulations, procedures, and oversight that have been instituted to safeguard the public and national security against the deliberate use of BSAT and addresses both physical security and personnel reliability. The committee was also asked to consider the impact of biosecurity policies and regulations on the ability of the scientific community to conduct BSAT research.[2]

GUIDING PRINCIPLES FOR SCIENCE AND SECURITY

In considering its task, the committee developed a set of principles that should guide how research with biological select agents and toxins should be viewed and conducted. These principles also provide the lens through which the committee addressed the specific concerns of laboratory security and personnel reliability.

1. Research on biological select agents and toxins is essential to the national interest.
2. Research with biological select agents and toxins introduces potential security and safety concerns.
3. The Select Agent Program should focus on those biological agents and toxins that might be used as biothreat agents.
4. Policies and practices for work with biological select agents and toxins should promote both science and security.
5. Not all laboratories and not all agents are the same.
6. Misuse of biological materials is taboo in every scientific community.

RECOMMENDATIONS[3]

Personnel Reliability

For those concerned about the security of laboratories conducting BSAT research, personnel issues are among the most difficult and controversial.

---

[1]The leadership of the National Interagency Biodefense Campus was also involved in requesting this study, which was conducted through a contract with the National Institute of Allergy and Infectious Diseases.
[2]See Box 1-2 for the full statement of task.
[3]The numbering of recommendations follows the order in the Executive Summary.

Personnel reliability programs incorporate *screening*, that is identifying whether or not someone should be eligible to have access to BSAT; *monitoring* employee behavior and performance; and *managing* the workplace to reduce the risk of an insider either carrying out theft or sabotage or acting to assist others.

Screening

Personnel screening seeks to identify individuals who may pose a potential security risk as early as possible, ideally prior to hiring. The proportion of the population of job candidates who represent true security risks is unknown, but likely to be very small. Efforts at screening for rare individuals or behaviors will therefore inevitably struggle with concerns about either failing to identify someone who has the disqualifying behavior or identifying someone as having disqualifying background or behavior when s/he does not. And the more one tries to avoid letting a security risk get through the screening, the more one increases the number of innocent individuals who will "fail" the test.

The Current Process   The current screening process to select individuals to work in facilities conducting BSAT research is based on the search for a set of disqualifying behaviors and activities that automatically and permanently deny a person access. This Security Risk Assessment (SRA) relies on the standard criminal, immigration, and terrorist databases maintained by the FBI and Department of Homeland Security (DHS) for these purposes and used to conduct routine suitability or security screening for other federal agencies. **The committee concluded that the databases being used in the SRA are consistent with current U.S. government practices in determining the eligibility of persons to have access to classified and proprietary information and sensitive sites and are adequate for assessing whether applicants possess disqualifying background/activities.**

The committee also considered several potential additions to the screening process. **The committee concluded that there was insufficient information to say that routine or random drug testing would significantly reduce the risk of an insider threat. The committee noted, however, that use of illegal drugs provides insight into a person's judgment and reliability, which are critical attributes for those with access to highly pathogenic infectious agents.** An obvious omission from the current SRA is querying an applicant's financial and credit history. At least some consideration of credit history is common in many sectors as part of pre-employment screening and is standard practice in federal security clearance and suitability investigations. In most cases, however, the issue is not one of an individual's level of debt per se, but whether spending patterns provide a means to assess judgment and reliability and possible vulnerabilities. **The committee concluded that the difficulties in establishing a meaningful baseline make adding credit or financial history to the SRA** screening process too challenging. In any event, signs of sudden, unexplained affluence or evidence of irresponsible financial behavior would be appropriate to consider as part of the process of monitoring employees' behavior, which is discussed below.

The committee also considered the issue of how determination of eligibility is made. In the current SRA, any discovery or admission of disqualifying factors or behavior automatically and permanently denies access for that individual. The current SRA system has no statute of limitations on disqualification: it does not matter how long ago the offense was committed. There is also no consideration of extenuating circumstances. The only appeal is to permit correction of factual errors. By contrast, information collected under other current federal suitability and security screening is subject to an adjudication process, whereby issues such as how long ago the offense occurred, whether recent behavior shows positive or negative trends, and mitigating circumstances are taken into account to determine whether to grant access to protected information. **The committee concluded that the questions raised about the current automatic and permanent disqualifications were sufficiently serious that it would be worthwhile to change the system to incorporate a broader appeal process more aligned with personnel security practices already in place across the government.**

These conclusions with regard to potential changes are conditional because the committee believes the appropriateness of additional measures, in some cases, depends on whether or not the select agent list is stratified, as recommended below.

RECOMMENDATION 5: The current Security Risk Assessment screening process should be maintained, but the appeal process should be expanded beyond the simple check for factual errors to include an opportunity to consider the circumstances surrounding otherwise disqualifying factors.

Identifying Potential Threats through Testing   Current policy discussions have included questions about whether to require more extensive testing and evaluation of applicants to work with BSAT materials, perhaps as part of a formal Personnel Reliability Program. Some government agencies and private entities, including academic institutions, have considered undertaking additional screening using psychological or psychophysiological tests.

At least two different types of problems need to be addressed when individuals are screened to identify those who potentially pose a threat. One arises in determining the normal range of adult personality; persons outside of this range would be identified as those who either might attempt deliberate deception or those who might be susceptible to corruption or recruitment to aid in the theft of materials or acts of sabotage. Another involves identifying individuals suffering from a range of serious personality disorders that might lead to their using BSAT materials to deliberately cause harm or assist others in doing so. Polygraph and integrity tests and personality assessment tools are among the instruments used to assess related factors and conduct screening.

The committee concluded that there is no "silver bullet," that is, no single assessment tool that can offer the prospect of effectively screening out every potential terrorist. Although it can be appropriate for organizations to employ integrity testing and clinical personality assessments as part of screening to serve other purposes, the committee reached the same conclusion concerning polygraph testing as was reached by a 2003 NRC committee that applies even more broadly, namely to its use in security screening: "Polygraph testing yields an unacceptable choice for...employee security screening between too many loyal employees falsely judged deceptive and too many major security threats left undetected. Its accuracy in distinguishing actual or potential security violators from innocent test takers is insufficient to justify reliance on its use in employee security screening in federal agencies."

Monitoring and Management to Achieve a Safe and Secure Research Environment

Once an individual is cleared by the SRA, certification is in effect for five years. However, the FBI continues to monitor cleared individuals using selected databases; the FBI also receives automatic notices in some instances, for example, when an individual is arrested and fingerprinted. But this process cannot be expected to address all disqualifying factors or, more importantly, all significant issues and personal changes that could occur in an individual's life during the five-year period of certification, including those that could potentially result in his or her becoming a security risk. Efforts to ensure personnel reliability will have to come from those laboratories where BSAT research is conducted and specifically from increased engagement by managers and staff.

A goal in any organization where safety is a central challenge should be to foster a culture where individuals watch out for each other and take responsibility for both their own performance and that of others. When this works well, the environment and culture reinforce a positive and inclusive ethic that promotes excellent performance. Security then becomes an additional goal, although many of the components of a safety-oriented culture serve security goals as well. A key component in a culture of trust and responsibility relevant to personnel reliability is a climate inducing self- and peer-reporting and providing mechanisms for such reporting. Management plays an essential role and has important responsibilities, not the least of which is to provide mechanisms for people to self-report problems and relay concerns about others via a safe mechanism (e.g., ombuds offices, hotlines, confidential reporting systems) and to enable individuals to obtain help in dealing with concerns proactively (e.g., employee assistance programs). Although often focused on safety concerns, these processes can serve security as well.

Research on preventing a wide range of types of insider threats suggests that, even in circumstances where one might assume an individual would attempt to conceal his or her malevolent intent in order to escape detection, *in many cases there will be signs or signals that something is wrong prior to an event.* Those cases in which an individual's action is genuinely spontaneous are rare. While no system can guarantee success in preventing an illegal act, the warning signs occur often enough that it is reasonable to believe that active, sustained monitoring and management could detect many of them and provide the basis for prevention. The research also suggests that training people to watch for and recognize the warning signs is essential and that, in the absence of such training, these signs are likely to be missed.

RECOMMENDATION 1: Laboratory leadership and the Select Agent Program should encourage and support the implementation of programs and practices aimed at fostering a culture of trust and responsibility within BSAT entities. These programs and practices should be designed to minimize potential security and safety risks by identifying and responding to potential personnel issues. These programs should have a number of common elements, tailored to reflect the diversity of facilities conducting BSAT research:

- Consideration should be given to including discussion of personnel monitoring during (1) the initial training required for all personnel prior to gaining access to BSAT materials and annual refresher updates and (2) safety inspections to obtain a more complete assessment of the laboratory's ability to provide a safe and secure research environment.
- More broadly, personnel with access to select agents and toxins should receive training in scientific ethics and dual-use research. Training should be designed to foster community responsibility and raise awareness of all personnel of available institutional support and medical resources.
- Federal agencies overseeing and sponsoring BSAT research and professional societies should provide educational and training resources to accomplish these goals.

Managing BSAT Research and the Select Agent Program

In addition to issues of personnel reliability, the committee addressed other issues related to physical security and to operation of the Select Agent Program, which led to several additional recommendations.

Facilitating Stakeholder Input by Formation of a BSAT Advisory Committee

One of the frequent themes that emerged from the public consultations held by the EO Working Group and the NSABB and in the committee's own public sessions and site visits was the need for increased and more systematic communication among those agencies funding BSAT research, those agencies administering the Select Agent Program, and those entities conducting BSAT research. The creation of the National Select Agent Registry as a single point of contact for agents regulated by both CDC and APHIS has been almost universally applauded for simplifying the regulatory environment and providing coordinated guidance. But because BSAT research is carried out and supported by several federal agencies, the committee believes a more formal structure is needed to engage the community of stakeholders in the continued operation of the program.

> **RECOMMENDATION 2: To provide continued engagement of stakeholders in oversight of the Select Agent Program, a Biological Select Agents and Toxins Advisory Committee (BSATAC) should be established. The members, who should be drawn from academic/research institutions and the private sector, should include microbiologists and other infectious disease researchers (including select agent researchers), directors of BSAT laboratories, and those with experience in biosecurity, animal care and use, compliance, biosafety, and operations. Representatives from the federal agencies with a responsibility for funding, conducting, or overseeing select agent research would serve in an ex officio capacity. Among the responsibilities of this advisory committee should be the following:**
>
> - **Promulgate guidance on the implementation of the Select Agent Program;**
> - **Facilitate exchange of information across institutions and sectors;**
> - **Promote sharing of successful practices across institutions and sectors;**
> - **Provide oversight for evaluation of the Select Agent Program;**
> - **Provide advice on composition/stratification of the list of select agents and toxins;**
> - **Convene regular meetings of key constituency groups; and**
> - **Promote harmonization of regulatory policies and practices.**

Stratification of the List of Select Agents and Toxins

The current list of select agents and toxins represents a diversity of pathogenic microorganisms and toxins with a wide range of potential for use as biothreat agents. Does this single list, all of which are subject to the same security procedures, represent the optimal solution?

The committee concluded that the present all-encompassing model for the list of select agents and toxins does not address appropriately the range of risks and vulnerabilities. Moreover, a list of more than 80 agents of varying risks dilutes attention from those that pose the greatest degree of concern, which may, in the process, render the nation less secure. It would be more effective to focus the highest scrutiny on those agents that are, indeed, of greatest concern and on those facilities with the equipment that enables weaponizing biological agents—and to offer a graded series of security procedures and policies for agents that pose less risk. For these reasons, the committee recommends reconsideration of both the purpose and composition of the list of select agents and toxins to reflect actual security concerns that merit inclusion on the list.

Although consideration of which specific agents and toxins should be on such a list is beyond the charge of the committee, we believe that stratification of the list of select agents and toxins is warranted. Stratification should be consistent with the original purpose of creating the list, namely to catalogue those agents posing a risk for use as a significant biothreat agent. Further, we believe that it is important to develop mechanisms for adding or removing agents from the list without unwarranted delay to ensure that the list remains reflective of legitimate concern. A procedure is needed to assess the risk posed by a biological agent that would initiate a formal process to add it to the list—or, equally important, to determine that an earlier estimation of threat has diminished and an agent should be taken off the list. Critical in consideration of adding or removing an agent from the list is the opportunity for significant information and input from external stakeholders, beyond the usual formal commenting process to government officials.

RECOMMENDATION 3: The list of select agents and toxins should be stratified in risk groups according to the potential use of the material as a biothreat agent, with regulatory requirements and procedures calibrated against such stratification. Importantly, mechanisms for timely inclusion or removal of an agent or toxin from the list are necessary and should be developed.

The BSATAC should advise the Select Agent Program on the implications that stratification of the list of select agents and toxins has on implementation of personnel screening, physical security requirements, and other procedures.

Accounting for Materials

It is prudent and appropriate for entities with the responsibility for BSAT laboratories to know what types of select agents and toxins are present in their facilities. In addition to maintaining records of materials in a facility for security purposes, such listings serve an important safety function in detailing materials of concern for laboratory personnel, as well as for first responders in emergencies. Current regulations provide highly specific guidance with respect to information to be collected. While the committee believes that it is useful and important to know which agents are present and where they are located, we question the value of measuring the quantity for living microorganisms, except for the amounts when acquired by a facility or transferred out to another facility. Because a new culture can be prepared with as little as a single microorganism, an individual would need only a miniscule—and undetectable—amount from a single vial to establish a new culture and grow up large volumes of the agent in a matter of hours or a day. Therefore, determining that the number of vials is the same from one moment to another provides no guarantee that agents have not been removed from the laboratory since the original number of vials or tubes could remain the same while the agent itself has been removed. **The committee, therefore, concluded that undue reliance on accounting practices, including counting vials, leads to false security and is counter-productive.**

> **RECOMMENDATION 4: Because biological agents have an ability to replicate, accountability is best achieved by controlling access to archived stocks and working materials. Requirements for counting the number of vials or other such measures of the quantity of biological select agents (other than when an agent is transported from one laboratory site to another) should not be employed because they are both unreliable and counter-productive, yielding a false sense of security. A registered entity should record the identity of all biological select agents and toxins within that entity, where such materials are stored, who has access and when that access is available, and the intended use(s) of the materials.**

It should be noted that this recommendation makes a distinction between select agents—which have the capacity to replicate—and toxins—which do not. This recommendation, therefore, does not change the requirement to keep records on the amount of a toxin but does recommend that inventories for both select agents and toxins should include information about who has access to these materials, when, and for what intended purpose.

With regard to another aspect of accounting for materials, **the committee concluded that, when specifically indicated by a risk assessment, a rule that "no one works alone"—defined as one person conducting work while being in direct communication with a second person who can affect a rescue—should be in place.** Since this is a safety measure with only indirect security benefits, security is best maintained by regulating access—namely, requiring log entry and exit systems and electronic identification cards for all personnel.

Security Based on Risk Assessments

Physical security is required of all facilities registered with the Select Agent Program. Each facility must develop and implement a written security plan, which is reviewed by either CDC or APHIS as part of the initial and ongoing facility registration process. Because each facility is different in design, different physical security methods are required to address site-specific security requirements. Determination of which physical security measures to include in a plan is made based on "a site-specific risk assessment and must provide graded protection in accordance with the risk of the select agent or toxin, given its intended use."

These select agent regulations provide overall guidelines for the content of site-specific security plans; however, they are sufficiently broad to allow for variation in their implementation. While this variation has benefits, it also creates inconsistencies and confusion as facility operators, contractors, and inspectors attempt to determine whether specific security measures at individual facilities sufficiently adhere to these guidelines. Moreover, many additional regulations have been separately imposed by different federal agencies, leading to inconsistencies in their application for a variety of reasons, in part because facilities and regulations differ. Addressing these inconsistencies and the problems they create would be highly beneficial both for security and cost-benefit, allowing cost-effective and consistent compliance with security needs and regulations.

> **RECOMMENDATION 6: The Select Agent Program should define minimum cross-agency physical security requirements, which recognize that facilities have unique risk-based security needs and associated design components, to assist facilities in meeting their regulatory obligations.**

The Select Agent Program can further assist institutions in interpreting physical security requirements by establishing a hotline or other mechanism for rapid response in answering questions about interpretation of the standards.

Evaluation

The committee believes that it is both appropriate and necessary to apply rigorous analytical methods to assess the mix of policies that promote both high-quality science and appropriate security. But assessing how and whether a program or programs achieve desired goals presents a particular evaluation challenge. If the policies are successful, nothing bad will happen. Following from the difficulty in assessing the effectiveness of programs that will be successful if there is no obvious effect—other than the absence of another action—it is likewise difficult to assess whether the various costs associated with the program are appropriate.

Independent evaluation can provide useful information on how the Select Agent Program is implemented and can identify important intended or unintended consequences of the program upon the research enterprise. The committee believes that new policies intended to improve physical security and personnel reliability should be carefully evaluated, along with the operation of the program overall. Relying on "dueling anecdotes" is not acceptable for establishing policy. The committee emphasizes that formal evaluation of the Select Agent Program is more than accumulation of metrics and demographic data.

> **RECOMMENDATION 7: Independent evaluation of the Select Agent Program should be undertaken to assess the relative benefits for achieving security, to consider the consequences of the program (intended and unintended) on the research enterprise, and to provide useful data about the Select Agent Program. Such evaluation, which may be coordinated through the BSAT Advisory Committee, should be provided with dedicated funding.**

Training of Inspectors

All select agent laboratories undergo regular inspections by CDC or APHIS, whether academic, commercial, or government and whether for research or public health. In addition to these inspections by agencies with statutory responsibility for the Select Agent Program, many funding agencies—including the Department of Defense and Department of Homeland Security—conduct their own inspections of research and facilities they support. Other federal agencies also may have responsibility for overseeing aspects of the facility and may conduct inspections. Finally, some state and local authorities inspect facilities within their jurisdiction.

Close coordination between CDC and APHIS in the Select Agent Program has served the research community well and should be expanded to include other government agencies with an involvement in BSAT research. Specifically, the committee encourages coordination and consolidation so that entities with select agent programs sponsored and/or regulated by different federal agencies are not subject to very different and possibly conflicting guidance and regulations or to duplicative inspections. In addition, it is critical to ensure that the requirements of multiple agencies are not contradictory; otherwise, the resulting confusion and uncertainty results in excess time and cost and increased difficulty of compliance.

Complaints about the nature of some inspections have arisen. Members of the community have cited the increasingly bureaucratic nature of some inspections, with expanding focus on the technical letter of the regulation without regard to the spirit of the regulation and its intended objective, and have expressed their concern that some inspectors have not had the technical knowledge needed to understand the specific nature of various risks. Much of this concern may stem from inspectors not sufficiently familiar with the nature of BSAT research. These challenges are even more severe for those government agencies that do not focus on select agent facilities but have a responsibility for inspecting them.

> **RECOMMENDATION 8: Inspectors of select agent laboratories should have scientific and laboratory knowledge and experience, as well as appropriate training in conducting inspections specific to BSAT research. Inspector training and practice should be harmonized across federal, state, local, and other agencies.**

Funding Facility and Compliance Costs

Security and compliance procedures called for under the Select Agent Program can be significant, with costs substantially higher than for similar laboratory facilities. Security guards, cameras, access card readers, biometric identification technologies, alarms, lockable freezers and incubators, and other security measures all add to the cost of operating a select agent laboratory.

Construction of secure laboratories where select agent research will be conducted is often funded by grants specific for that purpose. But select agent laboratories have significant ongoing security and safety sustainment costs that far exceed the indirect costs that grantee institutions receive to cover the costs of facilities, maintenance, and operations.

The implications of sustainable funding required to conduct select agent research are troubling. It is not acceptable, either for the institution or for safety and security, to diminish appropriate and necessary risk-based security procedures and resources, regardless of the availability of funding for the facility. The committee urges federal agencies that fund BSAT research to establish dedicated funding for ongoing security and compliance responsibilities associated with this type of research. This is an essential obligation, and no facility should operate without appropriate security measures in place. Although this type of funding structure may be unusual for biomedical research laboratories, it is not uncommon for funding those areas of science where central infrastructure plays an important role.

> **RECOMMENDATION 9: Because of considerable security and compliance costs associated with research on biological select agents and toxins, federal agencies funding BSAT research should establish a separate category of funding to ensure sustained support for facilities where such research is conducted.**

1

Introduction

THE PROMISE AND PERFORMANCE OF BSAT RESEARCH

More than 300 years ago, Antonie van Leeuwenhoek constructed a primitive microscope and made the first detailed descriptions of microorganisms. More than 200 years ago, Edward Jenner carried out the first experimental vaccination, using cow pox virus to build immunity in humans against the deadly smallpox virus. More than 100 years ago, Robert Koch isolated the *Bacillus anthracis* bacterium and postulated a causal relationship between specific microorganisms and disease.

From these early discoveries, scientists have built more than a century of research on microorganisms and infectious disease, including research on some of the most dangerous pathogens. Enormous advances have resulted in the development of vaccines and other treatments that have greatly diminished the risks posed by infectious disease agents. It is not an exaggeration to attribute increased human lifespan and better human health to the research of legions of microbiologists and other biomedical researchers on the biology of bacteria and viruses and the toxins they produce.

At the same time, these researchers have maintained safety and responsibility in the laboratory. Notwithstanding the enormous volume of infectious disease research that has been accomplished, there have been few incidents of pathogenic organisms being released into the environment by accident, negligence, or deliberate action. The incidence of laboratory-acquired infection is similarly exceedingly rare, even though many thousands of scientists handle highly pathogenic organisms daily. Moreover, scientists have become less tolerant of the possibility of release or accidental infection, working to improve biosafety as our understanding of biological materials and the risks they pose has increased. Scientific research is safer than it has ever been because of the increasing concern for safety and security and implementation of protective measures that minimize risk.

Among the larger group of pathogenic materials is a set of organisms and chemicals that pose not only a severe threat to the health of humans, plants, and animals, but also have the potential to be used deliberately to cause disease, prompt fear, or destroy agricultural or animal products. More than 80 of these most dangerous bacteria, viruses, toxins, and fungi have been officially listed as biological select agents and toxins (BSAT) and are subject to special security requirements.[1]

Whether deliberately deployed as a biological weapon or the result of a natural outbreak, the potential for mass human casualty or potentially catastrophic impact on plants or animals as a direct or indirect result of select agents is omnipresent. As the National Institute of Allergy and Infectious Diseases (NIAID) opened its most recent strategic plan for biodefense research, "biological weapons in the possession of hostile states or terrorists, as well as naturally occurring emerging and reemerging infectious diseases, are among the greatest security challenges to the United States" (NIAID 2007). The security and safety of our nation—as well as human and agricultural health around the world—depend upon a deep understanding of these organisms and toxins.

The most direct impact of research with BSAT is in the development of countermeasures against the agents themselves. Previous investment in research using what are now classified as select agents has yielded vaccines, drugs, and other treatments to combat agents such as smallpox, anthrax, and Ebola virus (Auchincloss 2007). Continuing efforts against these dangerous pathogens will improve our capacity to treat and prevent outbreaks when they occur, and advances in technology will enable more rapid detection of the presence of BSAT materials in the environment.

But the value of BSAT research is not limited to the development of medical countermeasures; in fact, greater understanding of BSAT materials will also enhance our ability to respond to a wide range of infectious diseases (NIAID 2008). What is learned about this small subset of pathogens can lead to strategies for responding to a much wider range of infectious diseases, extending the reach of BSAT research beyond the agents of acute concern to the much wider array of organisms with significant public health implications.

The nation's capacity to conduct research on BSAT materials has expanded significantly over the past several years. For example, the number of laboratories either in operation or under development that have the capacity to conduct research on the most dangerous pathogens—agents that pose the highest risk of life-threatening disease for which no vaccine or therapy is available, including

---

[1]These agents are defined in three sections of federal regulations: 42 CFR 73 for threats to "public health and safety," 7 CFR 331 for threats to "plant health or to plant products," and 9 CFR 121 for threats to "animal health or to animal products."

several select agents—has increased from two before 1990 to five before the terrorist attacks of 2001 to 15 or more that are operational or under development at the time of this report (GAO 2007).[2] Such laboratories are no longer limited to the federal government but now include facilities in academic institutions, state and local public health departments, and in the private sector. This expansion can be attributed to growing concerns about our limited understanding of dangerous pathogens, increasing ability to add to this understanding, and an influx of federal support for these activities. One large federally supported program highlights the growth as a result of increasing government support: since 2003, NIAID has supported the development of 11 Regional Centers of Excellence (RCEs) for Biodefense and Emerging Infectious Diseases,[3] which involve nearly 500 principal investigators (PIs)—most new to biodefense—at almost 300 institutions participating in RCE research activities (Concept Systems 2008). The laboratories provide a venue for work with potentially dangerous pathogens, including those on the list of select agents and toxins.

THE NATURE OF THE THREAT

BSAT materials have the potential for dramatic impact on human, plant, and animal health. For this reason, there is a growing concern that these agents may be used for intentional harm or to induce public panic. The anthrax attacks of 2001 are a prime example. In addition to killing five people and infecting 22, this attack had a dramatic impact on our nation and was estimated to have had a direct economic impact of more than $1 billion.

Clearly, there is genuine and legitimate concern that laboratories working with select agents and toxins should receive special security and safety attention that other types of biological research would not require. Even though many of the materials on the select agent list may be found in natural environments, some laboratories maintain purified strains of the most dangerous pathogens. In addition, laboratory workers not only have access to these materials but also may possess the technical knowledge of how to grow them in the laboratory, although not necessarily the technical knowledge needed to weaponize them.

This report therefore addresses policies and practices directed at securing those laboratory facilities in which work is done with BSAT materials. The intent is to protect the laboratories and the agents from threats posed by outsiders as well as insiders. Although the report does not focus on biosafety, some of the methods that prevent accidental infection or release also serve to enhance

---

[2]A 2009 Government Accountability Office (GAO) report lists seven operational labs as of 2009—four operated by the federal government, two by academic institutions, and one by a private nonprofit organization. GAO counts seven additional facilities in development, including one that will replace an existing facility (GAO 2009c).

[3]See <http://www3.niaid.nih.gov/LabsAndResources/resources/rce/introduction.htm> for more information about the RCEs.

security and may be discussed. But the focus of the report is on the security of the agents, facilities, and personnel.

There are specific issues concerning BSAT research that will be addressed in this report, but it is important first to consider the threat itself. What are the specific scenarios of concern and which eventualities are to be prevented? While a fully deliberative consideration of the threat is beyond the scope of this report, the committee selected several examples of possible threats as the context for discussion:

- A dedicated terrorist or criminal who may break into a BSAT laboratory with the intent to steal dangerous pathogens or to cause an intentional release.
- An individual working in a BSAT laboratory, with access to pathogens, who may take them out of the laboratory for improper use.
- An individual working in a BSAT laboratory who may serve as an accomplice or conduit for others wishing to do harm, whether deliberately or unwittingly.

Some individuals cited as examples above are motivated by ideology, while others are subject to pressure from the promise of money or other benefits; still others have their judgment compromised by a temporary or permanent condition or personal crisis. Although some security and personnel reliability strategies serve to address multiple threats, others are specific to a given population.

It is important to keep in mind that access to a pure culture of a select agent or toxin alone does not represent a major biothreat, although it can be more than sufficient for an act that is intended to evoke fear rather than mass casualty. To have widespread impact on health, the agent must be grown in reasonable quantity with technically complex facilities and specialized equipment, and may need to be stabilized to remain viable, packaged, delivered to a susceptible population, and dispersed in a method that allows the organisms to retain their virulence. Access to the starting material represents only the first and, in many cases, least sophisticated step in this process. Nonetheless, denying would-be terrorists ready access to BSAT materials is an important component of national security.

SUMMARY OF THE CURRENT SELECT AGENT PROGRAM[4]

After the anthrax attacks of 2001, the United States expanded the existing regulations governing the transfer of BSAT materials among laboratories

---

[4]Throughout this report, the term "Select Agent Program" is used to refer to the National Select Agent Registry Program, which oversees activities related to biological select agents and toxins.

registered with the Centers for Disease Control and Prevention (CDC) of the Department of Health and Human Services (HHS) and the Animal and Plant Health Inspection Service (APHIS) of the U.S. Department of Agriculture (USDA) into a rigorous formal oversight system to decide that persons seeking to possess, use, or transfer select agents or toxins have a lawful purpose. It also defined how facilities possessing BSAT materials would be protected. Appropriately defined, such a system would ensure that pathogens and toxins would be accessible only to legitimate researchers. The objective of material control for the life sciences focuses on methods to ensure that any individual with access to select agents would be trustworthy and that the agents would be secure within each facility housing BSAT materials. Chapter 2 describes in greater detail the current policy and regulatory framework governing BSAT research in the United States.

The scope of the Select Agent Program is circumscribed by those agents and toxins on the formal select agent list. CDC maintains the list for human pathogens, while APHIS maintains the list for plant and animal pathogens.[5] The list, first introduced in 1997, has grown from an initial 42 CDC agents and toxins to its current 82 CDC and USDA agents and toxins. The current list includes 40 HHS-only agents, 10 overlap agents, and 32 USDA-only agents (24 animal pathogens and 8 plant pathogens).[6] A formal process for determining whether an agent or toxin should be on the select agent list has been developed. In July 2009, for example, a notice in the *Federal Register* began the process of public comment on a proposal to add the SARS-associated coronavirus to the list (HHS 2009a), followed by a second notice in August 2009 of a proposal to add Chapare virus to the list (HHS 2009b). Regulations require a formal biennial review process during which the entire list is reviewed and agents or toxins may be added or removed.

The USA PATRIOT Act of 2001 (Public Law 107–56, October 26, 2001) established prohibitions on the possession of select agents by several categories of "restricted persons," including convicted felons or those who had received a dishonorable discharge from the U.S. military, foreign nationals from countries designated as supporting terrorism, and current users of illegal drugs. The Act also made it an offense for a person to knowingly possess any biological agent, toxin, or delivery system of a type or in a quantity that, under the circumstances, is not reasonably justified by prophylactic, protective, bona fide research, or other peaceful purpose.

The provisions of the USA PATRIOT Act were subsequently augmented

---

It is a joint activity of the U.S. Department of Agriculture's Animal and Plant Health Inspection Service and the Department of Health and Human Services' Centers for Disease Control and Prevention.

[5]A few BSAT materials that affect both humans and animals are considered "overlap agents" and appear on both the CDC and APHIS lists.

[6]See Table 2 2 for the current list of select agents and toxins.

by the Public Health Security and Bioterrorism Preparedness and Response Act, known as the Bioterrorism Act of 2002 (Public Law 107–188, June 12, 2002). This Act added requirements for regulations governing possession of select agents, including approval for laboratory personnel by the Attorney General following a background check by the Federal Bureau of Investigation (FBI). Entities possessing BSAT materials are required to register and have plans in place for ensuring: (1) the physical security of the BSAT materials in their possession; (2) appropriate biosafety to guard against an accident or an accidental release of BSAT materials; and (3) the ability to respond in the event that an accident, theft, or release did occur. Inspections by CDC and APHIS are used to assess adequacy of the plans; the two organizations also provide training and compliance assistance for those who are subject to the regulations. The Select Agent Program began operation with interim rules in 2003, and final rules were issued in April 2005, as three sections governing human, plant, and animal agents and toxins (HHS 42 CFR 73 (Humans); USDA 7 CFR 331 (Plants); and 9 CFR 121 (Animals)). APHIS and CDC work to ensure that their separate activities are coordinated and require the same types of policies, actions, and reporting from those they regulate.

RELATIONSHIP BETWEEN BIOSECURITY AND BIOSAFETY[7]

The concepts of "biosafety" and "biosecurity" are related and frequently complement one another, but they also differ in important ways. The fifth edition of the HHS manual *Biosafety in Microbiological and Biomedical Laboratories* (BMBL), which sets standards for how U.S. laboratories conduct research with biological agents and toxins, defines biosafety programs as those that "reduce or eliminate exposure of individuals and the environment to potentially hazardous biological agents," while the "objective of biosecurity is to prevent loss, theft or misuse of microorganisms, biological materials, and research-related information" (CDC/NIH 2007:105).[8] One frequently used description

---

[7]This section draws on the discussion in *Biosafety in Microbiological and Biomedical Laboratories*, 5th ed. (CDC/NIH 2007), Section VI, available at <http://www.cdc.gov/OD/ohs/biosfty/bmbl5/bmbl5toc.htm>.

[8]It should be noted that the use of the term "biosecurity" presents a number of difficulties. At its most basic, the term does not exist in some languages, or is identical with "biosafety"; French, German, Russian, and Chinese are all examples of this immediate practical problem. Even more serious, the term is already used to refer to several other major international issues. For example, to many "biosecurity" refers to the obligations undertaken by states adhering to the Convention on Biodiversity and particularly the Cartagena Protocol on Biosafety, which is intended to protect biological diversity from the potential risks posed by living modified organisms resulting from modern biotechnology. (Further information on the Convention may be found at <http://www.cbd.int/convention/> and on the Protocol at <http://www.cbd.int/biosafety/>.) "Biosecurity" has also been narrowly applied to efforts to increase the security of dangerous pathogens, either in the laboratory or in dedicated collections; guidelines from both the World Health Organization (WHO

of the difference offers a quick and accessible explanation: "Biosafety is about protecting people from bad 'bugs'; biosecurity is about protecting 'bugs' from bad people."

As discussed in the BMBL, the systems developed for biosafety and biosecurity have a number of common elements:

> Both are based upon risk assessment and management methodology; personnel expertise and responsibility; control and accountability for research materials including microorganisms and culture stocks; access control elements, material transfer documentation, training, emergency planning, and program management. . . . Biosafety looks at appropriate laboratory procedures and practices necessary to prevent exposures and occupationally acquired infections, while biosecurity addresses procedures and practices to ensure that biological materials and relevant sensitive information remain secure. Both programs assess personnel qualifications. . . . Both programs must engage laboratory personnel in the development of practices and procedures that fulfill the biosafety and biosecurity program objectives but that do not hinder research or clinical/diagnostic activities. The success of both of these programs hinges on a laboratory culture that understands and accepts the rationale for biosafety and biosecurity programs and the corresponding management oversight. (CDC/NIH 2007:105-106)

Not all aspects of biosafety and biosecurity are compatible. One widely used example is the kind of signs each would dictate for display in a laboratory. For biosafety purposes, good practice would require having a sign on the *outside* of the laboratory door to alert people that work was going on with a potentially dangerous pathogen; the information would include the name of the agent, any specific hazards, and contact information for the researcher. From a security point of view, displaying this kind of information would only make the task of a would-be thief or saboteur easier. Sharing information about the type of research being carried out and the safety practices in place in a laboratory in the name of open communication and public trust with the surrounding community might also arouse the concern of security professionals who would prefer to see more restricted use of such information.[9]

2004) and the Organization for Economic Cooperation and Development (OECD 2007) use this more restricted meaning of the term. In an agricultural context, the term refers to efforts to exclude the introduction of plant or animal pathogens. (See NRC 2009a:8-9 for a discussion of this and other issues related to terminology.) Earlier NRC reports (2004ab, 2006, 2009ab) confine the use of "biosecurity" to policies and practices to reduce the risk that the knowledge, tools, and techniques resulting from research would be used for malevolent purposes. The BMBL and this report use the term to cover security for both pathogens and for the information that results from research.

[9]To that end, some select agent laboratories do not broadcast their location, even if that information is considered public. Of course, emergency services and law enforcement are aware of the location and operation of these labs.

Despite these types of differences, good biosafety practices can provide an essential foundation for biosecurity.[10] But biosafety alone will not provide all of the aspects of good biosecurity, which must also address the risks posed by those with malevolent intent. Responding to these risks while also enabling a vigorous and productive research environment is the challenge to which this report attempts to respond.

THE IMPACT ON SCIENTIFIC RESEARCH

While the scientific community is vitally concerned about security threats posed by BSAT research, it is also cognizant of the possible unintended consequences on the scientific community by overzealous application of policies and procedures implemented in the name of enhanced security. If procedures are beyond that necessary to address the risk, the unintended consequence may be that top scientists are dissuaded from engaging in BSAT research, perhaps especially younger researchers. This will threaten the security of the nation because knowledge of pathogens and the public health measures to protect against them will be diminished. Therefore, the challenge for BSAT research is to implement those measures that promote security and simultaneously facilitate scientific progress in research. Similar concerns apply to others subject to the select agent regulations, including public health professionals who are essential to the nation's response to biological emergencies.

Future discoveries and successful research on select agents specifically— and in the life sciences more generally—depend on a healthy, vibrant, and sustainable research environment. Scientific progress requires that the best and most creative researchers be encouraged to seek out and solve interesting and important problems. This, in turn, requires minimizing the amount of unnecessary regulation and burdensome recordkeeping, which serve as impediments, and providing clear justification and transparency regarding those adopted for legitimate reasons, such as enhancing security.

Science is characterized by the free flow of information and the ability of research scientists to pursue lines of investigation that yield the most promising results. Publishing is the coin of the realm in science, and life scientists conduct research that is published in many thousands of peer-reviewed journals. The vast majority of research—including research with select agents and toxins—is not classified and not subject to restriction with respect to publication. Open exchange of ideas is essential because it encourages researchers to pursue research questions in a given area of science, and it allows scientists to share their research findings and follow new directions wherever they lead.

For years, when a research project has raised safety or ethical concerns, the

---

[10]This may be particularly important in developing countries, where improving biosafety can also bring many other benefits.

work has been subjected to oversight. In most cases, this has been local to the institution, with committees designated by the federal government to carry out the review. These review committees usually comprise scientific peers and those from other fields of study who have appropriate expertise, as well as representatives of the public (see Box 1-1 for examples of oversight committees operated at the institutional level).

Where safety concerns extend beyond the institution, national-level bodies provide oversight that is consistent across the country. Most prominent in basic research is the Recombinant DNA Advisory Committee (RAC), established by the National Institutes of Health (NIH) in 1974 in response to public concerns

---

BOX 1-1
Scientific Oversight Committees

Several areas of research are monitored by scientific oversight committees. In each case, these committees operate at the level of individual institutions, providing oversight for research conducted at that institution. In most cases, however, they are designated by the federal government, allowing a mix of institutional decisions with national reporting and allowing one institution to rely upon decisions made by similar committees at other institutions. These oversight committees comprise scientific peers along with experts in other appropriate fields of study as well as members of the public.

Institutional Review Boards (IRBs) are charged with protecting the rights and welfare of human research subjects recruited to participate in research activities. IRBs are required to register with HHS' Office of Human Research Protections.

Institutional Biosafety Committees (IBCs) are charged with reviewing research involving recombinant DNA, although many IBCs have chosen to review other forms of research that involve potential biohazards—including some BSAT research. IBCs are required to register with NIH's Office of Biotechnology Activities.

Institutional Animal Care and Use Committees (IACUCs) are charged with ensuring the appropriate care and use of all animals involved in research, training, and biological testing by overseeing an institution's animal program, facilities, and procedures. The existence of IACUCs is stipulated in the Animal Welfare Act.

Embryonic Stem Cell Research Oversight (ESCRO) Committees are recommended by the National Academies' *Guidelines for Human Embryonic Stem Cell Research* to provide ethical oversight on the field of human embryonic stem (hES) cell research (NRC/IOM 2005, 2007, 2008). ESCRO committees oversee all issues related to derivation and use of hES cell lines, review and approve the scientific merit of research protocols, review compliance with relevant regulations, maintain registries of hES cell research, and facilitate education of hES cell researchers.

about manipulation of genetic materials and use of recombinant DNA technology. The RAC developed and maintains the *NIH Guidelines for Research Involving Recombinant DNA Molecules*, which has become the standard of safe scientific practice in the use of recombinant DNA. It also considers other matters relevant to recombinant DNA, including the review of human gene transfer trials or novel protocols that raise new scientific, safety, or ethical considerations. Even though the RAC is a federally chartered committee, its members are drawn from the extramural scientific community, i.e., outside NIH.

Voluntary guidelines can have a significant impact within the scientific community. For example, there is no legislation mandating the use of the BMBL (CDC/NIH 2007), yet these guidelines are almost universally followed.[11] And the National Academies' *Guidelines for Human Embryonic Stem Cell Research* (NRC/IOM 2005, 2007, 2008) has been adopted nationwide, even without federal legal standing.[12]

Unlike these models, the oversight and screening structures for the Select Agent Program are considerably more substantial than those applicable to other biological research, especially in the involvement of outside oversight bodies and groups not necessarily composed of scientific peers. Moreover, BSAT research is the only area of biological research that requires verification of personnel beyond assessment of technical competence to carry out the proposed research protocols. For non-BSAT research, there has been no issue as to whether the individual may be trusted not to do harm. Thus, select agent regulations will be unfamiliar to most scientists. Many researchers also may find the regulations to be a significant and unusual burden. Members of the community have expressed concern about the potential impact of the regulations on recruiting and retaining scientists for select agent research—as well as public health professionals for detecting and responding to biological emergencies (e.g., HHS 2005; ABSA 2009; FASEB/AAMC 2009).

Scientific careers often involve protracted mentored training, not to mention the pressure to produce publishable findings. With biomedical researchers now on average well into their 40s before receiving their first independent research grant (NRC 2005), severe disincentives to pursue research careers already exist, and an additional burden placed on those who pursue research on select agents further challenges their decision to follow a career that involves select agents and toxins. During a visit to the New England Regional Center of Excellence, for example, the committee learned that a significant majority of

---

[11] Although several parts of the BMBL are used by CDC and APHIS in enforcing the select agent regulations, the guidelines are utilized much more widely than these required elements, including in laboratories outside of the United States.

[12] Even though the National Academies' *Guidelines* have no federal standing, several states have incorporated aspects into state-level, legally binding regulations, and some research sponsors similarly require compliance for their grantees.

graduate students who began the clearance process to work with select agents did not complete all steps necessary.[13]

Finally, the committee has learned of a number of researchers at several institutions who chose to destroy their inventories of select agents and toxins rather than incur the cost and inconvenience of the security requirements and personnel screening of the Select Agent Program (e.g., Wilkie 2004). These research scientists chose to pursue other interesting research questions, rather than go through the arduous task posed by working with select agents and toxins.

THE CONTEXT FOR THE CURRENT PROJECT: EXECUTIVE ORDER 13486

Concerns about whether the regulations in place for BSAT research in U.S. laboratories were adequate to address the risks of theft, misuse, or diversion of materials increased after the FBI announced in August 2008 that it had concluded that Bruce Ivins, a research scientist at the U.S. Army Medical Research Institute of Infectious Diseases (USAMRIID), was the perpetrator of the anthrax letter attacks in October 2001 (FBI 2008). There were other concerns from Congress and elsewhere about whether the "proliferation" of high containment laboratories as part of the increased funding for biodefense research after 2001 was increasing the risks of laboratory accidents as well as providing more targets for those who could pose security threats from either outside or inside the facilities (e.g., GAO 2007, 2008). The release of the report of the Congressionally chartered Commission on the Prevention of Weapons of Mass Destruction Proliferation and Terrorism, chaired by former senators Bob Graham and Jim Talent, drew additional attention to the perceived risks. The Commission's report, *World at Risk*, began its Executive Summary with the ominous conclusion that:

> ...unless the world community acts decisively and with great urgency, it is more likely than not that a weapon of mass destruction will be used in a terrorist attack somewhere in the world by the end of 2013. The Commission further believes that terrorists are more likely to be able to obtain and use a biological weapon than a nuclear weapon. The Commission believes that the U.S. government needs to move more aggressively to limit the proliferation of biological weapons and reduce the prospect of a bioterror attack. (WMD Commission 2008:xv)

---

[13] According to representatives of the laboratory, only about two of the 20 students who began the training and clearance process completed it.

The Commission recommended a number of steps to increase security at all U.S. high containment laboratories, not just those conducting BSAT research.

As part of the response to the various calls for increased regulation of high containment laboratories and BSAT research, President George W. Bush issued Executive Order (EO) 13486, *Strengthening Laboratory Biosecurity in the United States*, on January 9, 2009 (White House 2009). The EO established an interagency Working Group on Strengthening the Biosecurity of the United States, charged with conducting a comprehensive assessment of the efficiency and effectiveness of all laws, regulations, guidance, and practices related to physical, facility, and personnel security and assurance for BSAT research. The heart of the group's report, submitted to the President within 180 days (i.e., by July 9, 2009), would provide "recommendations for any new legislation, regulations, guidance, or practices for security and personnel assurance for all Federal and nonfederal facilities … and options for establishing oversight mechanisms to ensure a baseline standard is consistently applied for all physical, facility, and personnel security and assurance laws, regulations, and guidance at all Federal and nonfederal facilities…" (White House 2009). These recommendations would be supplemented by another extensive interagency review of biosafety practices being conducted by the Trans-Federal Task Force on Optimizing Biosafety and Biocontainment Oversight.[14]

To provide additional input, the Homeland Security Council staff at the White House requested two other studies. The first, which focused only on personnel reliability, was carried out by the National Science Advisory Board for Biosecurity (NSABB) and issued in May 2009 (NSABB 2009). The second study was requested from the National Research Council (NRC), resulting in this report.[15]

In addition to the reports that provide formal input into Executive Branch deliberations, a number of other relevant reports have been issued in recent months. The Defense Science Board (DSB) released a report in May 2009, focused on the *Department of Defense Biological Safety and Surety Program* (DSB 2009). The Defense Health Board (DHB) issued a report in April 2009 that addressed whether the military services needed to own and operate their own biodefense infrastructure and research program, which affects whether and how it carries out physical security and personnel reliability programs (DHB 2009). Two workshops on education by the American Association for the Advancement of Science (AAAS), one on so-called dual use education (AAAS 2008) and one focused on biosafety training (AAAS 2009), offered a

---

[14]Further information about the Trans-Federal Task Force, including a copy of its report (Trans-Federal Task Force 2009), may be found at <http://www.hhs.gov/aspr/omsph/biosafetytaskforce/>.

[15]The leadership of the National Interagency Biodefense Campus was also involved in requesting this study, which was conducted through a contract with NIAID.

number of findings and recommendations related to how the training could support personnel reliability.[16] These reports, as well as numerous meetings and discussions, have contributed to a lively and sometimes heated discussion of appropriate approaches to optimizing the security and the quality of BSAT research.

CHARGE TO THE COMMITTEE

The NRC appointed a committee with a broad range of expertise to carry out its statement of task, which is reproduced in Box 1-2 (short biographies of the committee members and project staff are contained in Appendix A).

The committee focused its attention on the environments in which BSAT research is conducted, which are a subset of the facilities cleared to work with select agents and toxins. While other entities such as state and local public health laboratories are subject to the select agent regulations, most do not have research as their primary focus.

The committee carried out its work over approximately 3½ months, with two in-person meetings and several site visits, as well as conference calls to begin and conclude its work. A list of the meetings and site visits, including the briefings received by the committee, are contained in Appendix B. The committee considered not only the experiences of select agent laboratories, but also related experiences in other sectors including nuclear power plants, academic nuclear research reactors, and the aviation industry—all of which have been concerned about personnel reliability for some time.

In the end, time constraints meant that the committee could not give equal attention to all elements of its task. Therefore the committee decided to concentrate on a set of issues that it believes are the most important, most critical, and most effective for both providing security and enabling the highest quality research to be carried out in an environment that can attract and retain the best scientists. The focus of the report was also informed by the elements that had prompted the greatest amount of discussion within the scientific community and at the public consultations organized by the NSABB and EO Working Group. There are two other items that elicited significant interest in the public consultations but that could not be considered in this report. The committee believes these are essential to the safe conduct of BSAT research, but time did not allow a thorough review and assessment:

- *Transportation of Select Agents* Some have identified transportation of select agents and toxins as the weak link in security procedures. Agents taken from one highly secure facility to another may be at risk for

---

[16]The term "dual use" refers to research that, although carried out for beneficial purposes, could yield knowledge, tools, or techniques with the potential to be misused to cause deliberate harm.

BOX 1-2
Committee Statement of Task

An ad hoc committee will assess the efficacy of regulations, procedures, and oversight that have been instituted to safeguard the public and national security against the deliberate use of biological select agents and toxins (BSAT). The assessment will specifically take into account programs for laboratory security to protect against external threats and, in particular, personnel reliability assurance programs (protection against internal threats). The committee will not address biosafety (protection against accidental releases) except to the extent that biosecurity impinges on biosafety measures. The committee will also assess the impact of biosecurity policies and regulations on the ability of the scientific community to conduct BSAT research. The committee will evaluate progress since 2001 and identify opportunities for the U.S. government to optimize the balance between controlling and mitigating security risks associated with BSAT research and ensuring the benefits of BSAT research for force and public health protection. The committee's conclusions and recommendations will be designed to inform policy discussions in the United States regarding necessary steps to balance the security risks and benefits of BSAT research and to harmonize policies across the government, including government-funded extramural research.

Based on expert knowledge of the current oversight systems for BSAT research, information gathered in the course of the study about the specifics of the programs that have been implemented by each of the federal agencies with active BSAT research programs, and information about personnel assurance programs outside the federal government that might offer useful models or practices, the committee should:

---

theft during transportation because security during this process may be minimal. In addition, the physical security solutions and workforce involved in transporting select agents may not adhere to the requirements for select agent facilities (described in Chapter 2).

The committee did not have the time to fully explore this issue, especially because shipping requirements are based upon international standards regarding the transportation of hazardous materials. Therefore, any changes to transportation procedures for select agents could have unintended consequences for shipping of other materials and could unintentionally complicate the international exchange of biological materials.

- *Cybersecurity* Because many of the physical security solutions depend on technology—such as cameras, electronic access cards, electronic inventory systems—there is a risk posed by those individuals able to hack into these command and control systems. To the extent that these systems may not be fully secure, additional risks exist.

1. Develop a set of principles and questions to be addressed in developing a framework to guide programs that provide and oversee laboratory security and personnel reliability systems for BSAT research. This framework should optimize benefits, minimize risk, and facilitate the productivity of research.

2. Review and assess the efficacy and cost/benefit of similar laboratory security, personnel reliability, and BSAT accountability programs of federal agencies to explore best practices across the federal government. The review should consider the implementation of existing legislation, regulations, guidance, policies, and practices as they relate to both federal laboratories or programs and research facilities at representative extramural laboratories funded by government programs.

   The assessment will include potential impacts on the ability to attract and sustain quality scientists to conduct research on BSAT and identification of factors responsible for barriers to research on DOAT in the extramural environment.

   The committee will make recommendations for refining existing programs and procedures affecting both intramural and extramural facilities that will achieve greater productivity in research objectives, optimize management to reduce risk, and produce improved uniformity, transparency, and efficiency in research on BSAT.

3. Make recommendations to inform policy decisions for achieving an effective system for oversight to ensure compliance with these programs and procedures.

The committee may consider examples of facility security and personnel assurance programs in other settings, including those from outside the BSAT domain and those outside the federal government, that might offer lessons or best practices.

The committee's conclusions and recommendations were developed independent of the other reports on these topics including those identified above, although the committee did have access to those reports that had been released while the committee was engaged in its work (AAAS 2008, 2009; DHB 2009; DSB 2009; NSABB 2009). The committee did not have access to the reports from the EO Working Group or the Trans-Federal Task Force, which had not been released before the completion of this report.

ORGANIZATION OF THE REPORT

After the brief introduction to the issues addressed in the report in this chapter, Chapter 2 contains basic factual material describing the current regulatory environment including the development and operation of current U.S. policies to govern BSAT research, a review of other federal regulations related to BSAT research, and a brief discussion of how BSAT research is regulated in other countries. Chapter 3 sets out some basic principles that guided the committee in selecting those issues it chose to emphasize and in reaching its conclusions and recommendations. Chapters 4 and 5 discuss specific issues and offer the committee's analysis and assessments, including its conclusions and recommendations.

2

The Current Regulatory Environment

INTRODUCTION

Fundamental International Commitments

The fundamental international commitments not to use disease as a weapon are embodied in the Geneva Protocol, which was signed in 1925 and entered into force in 1928, and the Biological and Toxin Weapons Convention (BWC), which was signed in 1972 and entered into force in 1975.[1] The Geneva Protocol prohibits the first use of chemical and biological warfare but does not ban production, storage, or transfer. That gap was closed by the BWC and later by the Chemical Weapons Convention of 1993. Article I of the BWC states:

Each State Party to this Convention undertakes never in any circumstances to develop, produce, stockpile or otherwise acquire or retain:

(1) Microbial or other biological agents, or toxins whatever their origin or method of production, of types and in quantities that have no justification for prophylactic, protective or other peaceful purposes;

(2) Weapons, equipment or means of delivery designed to use such agents or toxins for hostile purposes or in armed conflict.

---

[1]The Geneva Protocol's formal title is the Protocol for the Prohibition of the Use in War of Asphyxiating, Poisonous or Other Gases, and of Bacteriological Methods of Warfare and the BWC treaty's formal title is the Convention on the Prohibition of the Development, Production and Stockpiling of Bacteriological (Biological) and Toxin Weapons and on Their Destruction. The United States signed the Geneva Protocol in 1925 but did not ratify it until 1975, at the same time the Senate ratified the BWC.

The BWC does not prohibit research on defenses against biological weapons, which a number of countries, including the United States and its major allies, have continued.

Of more direct relevance to bioterrorism, United Nations (UN) Security Council Resolution (UNSCR) 1540, passed in 2004 with strong support from the United States, imposes a binding international commitment on all UN members not to provide "any form of support to non-State actors that attempt to develop, acquire, manufacture, possess, transport, transfer or use nuclear, chemical or biological weapons" (UN 2004). UN member states must undertake and enforce domestic measures against the proliferation of weapons of mass destruction (WMD), related materials, and the means to deliver them, including specific measures such as effective border controls and physical security. "If implemented successfully, each state's actions will significantly strengthen the international standards relating to the export of sensitive items and support for proliferators (including financing) and ensure that non-state actors, including terrorist and black-market networks, do not gain access to chemical, nuclear or biological weapons, their means of delivery or related materials" (Department of State 2009a).[2]

Evolution of U.S. Policies and Procedures for Research with Biological Agents and Toxins

Measures to Address Safety: Biosafety Guidelines

Over time, scientists have developed best practices for research with potentially dangerous biological agents or toxins—including but not limited to biological *select* agents and toxins (BSAT). Such practices are designed to ensure this research does not cause harm to those working in laboratories or to the broader public and environment because of accidents or accidental releases. In the United States, the National Institutes of Health (NIH) published its first edition of *Biosafety in Microbiological and Biomedical Laboratories* (BMBL) in 1984; the fifth edition was published in 2007 (CDC/NIH 2007).[3] The BMBL categorizes infectious agents and laboratory activities into four biosafety levels (BSL-1 through BSL-4) and establishes safety requirements for each level based on risk.[4]

---

[2]UNSCR 1810, passed in 2008, extended the mandate of UNSCR 1540 for an additional three years and urges states to complete its implementation (UN 2008).

[3]The World Health Organization (WHO) also produces a *Laboratory Biosafety Manual*. The first edition was published in 1983, and the third was released in 2004.

[4]The risk groups defined by the *NIH Guidelines for Research Involving Recombinant DNA Molecules* provide another classification of agents and toxins based on risk. See Box 5-1 for more information.

- BSL-1 laboratories are for working with agents and toxins that do not consistently cause disease in healthy human adults;
- BSL-2 laboratories are for working with agents and toxins that can be spread through puncture, absorption through mucus membranes, or ingestion;
- BSL-3 laboratories are for working with agents and toxins that are capable of aerosol transmission and that may cause serious or lethal infection; and
- BSL-4 laboratories are for working with agents or toxins that pose a high risk of life threatening disease that may be aerosol transmitted and for which there is no available therapy or vaccine.

BSL-3 and BSL-4 laboratories are considered "high" and "maximum containment," respectively. They require specialized expertise to design, construct, operate, and maintain. It should be noted that there are no inherent security requirements associated with the BSL levels; these are intended for safety. Security considerations are tied to the agents being used; the specific type of agent or toxin drives the requirement for security.

While some research and testing for the development of countermeasures for select agents can be conducted at BSL-2, work on the most dangerous agents requires BSL-3 and -4 conditions. High and maximum containment laboratories may also be necessary for some diagnostic and analytical services and for basic research on pathogenesis and other aspects of hazardous infectious agents. Table 2-1 contains a more detailed summary of the recommended practices, safety equipment, and facilities for each of the biosafety levels for infectious agents.

The BMBL guidelines have become the accepted practice in U.S. laboratories and have provided a model for similar practices in other countries. They are not codified in formal regulations, but are powerful performance-based standards for how laboratories are expected to operate. "According to federal grant policy and standard contract language, researchers and laboratory workers at institutions receiving federal funds are to be trained in the procedures described in the BMBL before they gain access to the laboratory" (Gottron and Shea 2009:6).

For the first time, the fifth edition of the BMBL contains a discussion of "biosecurity," whose objective is defined as preventing "loss, theft or misuse of microorganisms, biological materials, and research-related information" (CDC/NIH 2007:105).[5] The manual provides guidelines for a biosecurity program, including physical security and personnel management measures.

---

[5]The latest edition of the WHO manual also includes information on "laboratory biosecurity," which is defined as the "institutional and personal security measures designed to prevent the loss, theft, misuse, diversion or intentional release of pathogens and toxins" (WHO 2004:47).

TABLE 2-1 Recommended Practices, Safety Equipment, and Facilities for Biosafety Levels 1-4

BSL	Agents	Practices	Primary Barriers and Safety Equipment	Facilities (Secondary Barriers)
1	Not known to consistently cause diseases in healthy adults	Standard microbiological practices	None required	Laboratory bench and sink required
2	Agents associated with human disease Routes of transmission include percutaneous injury, ingestion, mucous membrane exposure	BSL-1 practices plus: • Limited access • Biohazard warning signs • "Sharps" precautions • Biosafety manual defining any needed waste decontamination or medical surveillance policies	Primary barriers: Class I or II biosafety cabinets (BSCs) or other physical containment devices used for all manipulations of agents that cause splashes or aerosols of infectious materials Personal protective equipment (PPEs): Laboratory coats, gloves, face protection as needed	BSL-1 plus: • Autoclave available
3	Indigenous or exotic agents with potential for aerosol transmission Disease may have serious lethal consequences	BSL-2 practices plus: • Controlled access • Decontamination of all waste • Decontamination of laboratory clothing before laundering • Baseline serum	Primary barriers: Class I or II BSCs or other physical containment devices used for open manipulation of agents PPEs: Protective laboratory clothing, gloves, respiratory protection as needed	BSL-2 plus: • Physical separation from access corridors • Self-closing, double-door access • Exhaust air not recirculated • Negative airflow into laboratory
4	Dangerous/exotic agents pose high risk of life threatening disease Aerosol-transmitted laboratory infections have occurred; related agents with unknown risk of transmission	BSL-3 practices plus: • Clothing change before entering • Shower on exit • All material decontaminated upon exit from facility	Primary barriers: All procedures conducted in Class III BSCs or Class I or II BSCs in combination with full-body, air-supplied, positive pressure personnel suit	BSL-3 plus: • Separate building or isolated zone • Dedicated supply and exhaust, vacuum, and decontamination systems • Additional requirements

SOURCE: CDC/NIH 2007.

Measures to Address Security

The BWC calls on its member states to enact legislation to support the implementation of the treaty. The United States passed the Biological Weapons Anti-Terrorism Act of 1989 (Public Law 101–298, May 22, 1990), which established penalties for violating the Convention's prohibitions, unless "(1) such biological agent, toxin, or delivery system is for a prophylactic, protective, or other peaceful purpose; and (2) such biological agent, toxin, or delivery system, is of a type and quantity reasonable for that purpose." In keeping with the treaty, the legislation focused on the *purpose* for which agents or toxins were possessed, rather than the agents themselves. The law authorizes the government to apply for a warrant to seize any biological agent, toxin, or delivery system that has no apparent justification for peaceful purposes, but prosecution under the law would require the government to prove that an individual did not have peaceful intentions (Atlas 1999). Since President Richard M. Nixon had already ended the U.S. offensive biological weapons program 20 years earlier,[6] the law attracted little attention when it was enacted.

THE BEGINNINGS OF THE SELECT AGENT PROGRAM

The first legislation that sought to limit the threat that biological agents or toxins from legitimate U.S. research laboratories would fall into the hands of terrorists was the Antiterrorism and Effective Death Penalty Act of 1996 (Public Law 104–132, April 24, 1996). The Act was passed amid rising concerns about terrorism, including with nuclear, chemical, or biological materials, in the wake of the 1993 World Trade Center and 1995 Oklahoma City bombings and the revelation of the Aum Shinrikyo's efforts to create biological as well as chemical weapons after its release of sarin in the Tokyo subway system. The immediate cause of the legislation was the attempt by a U.S. scientist with ties to white supremacist organizations to obtain plague-causing bacteria for potentially nefarious purposes (Stern 2000; Carus 2001; Gronvall 2008).

The Act contained several findings about the risks of bioterrorism:

(1) certain biological agents have the potential to pose a severe threat to public health and safety;

---

[6]On November 25, 1969, President Nixon issued National Security Decision Memorandum 35, which renounced the "use of lethal methods of bacteriological/biological warfare. The United States bacteriological/biological programs will be confined to research and development for defensive purposes (immunization, safety measures, et cetera)" (NSC 1969:2-3). The Memorandum stated, "This does not preclude research into those offensive aspects of bacteriological/biological agents necessary to determine what defensive measures are required" (NSC 1969:3). This order did not regulate *possession* of potentially dangerous pathogens, but instead focused on the *purpose* of the research. The BWC took a similar approach, as already noted.

(2)   such biological agents can be used as weapons by individuals or organizations for the purpose of domestic or international terrorism or for other criminal purposes;

(3)   the transfer and possession of potentially hazardous biological agents should be regulated to protect public health and safety; and

(4)   efforts to protect the public from exposure to such agents should ensure that individuals and groups with legitimate objectives continue to have access to such agents for clinical and research purposes. (Public Law 104–132, April 24, 1996, Sec. 511)

The Act required the Secretary of Health and Human Services (HHS) to issue regulations to govern the transport of biological agents with the potential to pose a severe threat to public health and safety through their use in bioterrorism. In establishing the list of materials to regulate, the Secretary was to consider: "(I) the effect on human health of exposure to the agent; (II) the degree of contagiousness of the agent and the methods by which the agent is transferred to humans; (III) the availability and effectiveness of immunizations to prevent and treatments for any illness resulting from infection by the agent; and (IV) any other criteria that the Secretary considers appropriate" (Public Law 104–132, April 24, 1996, Sec. 511). The Secretary delegated the authority to regulate these "select agents" to the Centers for Disease Control and Prevention (CDC). To ensure that the transfer of these agents was carried out only by and between responsible parties, CDC required that laboratories transferring select agents be registered and report each transfer.[7]

The initial list of select agents, introduced in 1997, contained 42 agents and toxins. It included some agents that could affect both humans and animals (for example, *Bacillus anthracis* and *Francisella tularensis*), but did not include those affecting only animals and plants. In drawing up the list, groups of experts from inside and outside the government focused on the agents and toxins that had been weaponized in the United States. and other offensive biological weapons

---

[7]"The purpose of registration was to control domestic transfers based upon a permitting system. A registered laboratory could legally transfer select agents only to another registered laboratory; some transfers were denied because of concerns about the adequacy of the facility proposed to receive the agent. Transfers to nonregistered laboratories were prohibited. Registration, however, was principally a matter of notification: a laboratory was obligated to notify relevant authorities of a transfer to another registered facility and that the transfer itself complied with applicable safety standards. Specific information about particular pathogens that the facility possessed did not have to be reported, not even if they were the subjects of extensive research, so long as they were not transferred. This was not intended to be a strict licensing system but merely a way of overseeing transfers and shipments of lethal pathogens" (NRC 2004a:75).

programs prior to the advent of the BWC or those that were considered to have the greatest potential for weaponization.[8]

In the wake of the terrorist attacks of September 11, 2001, and the *Bacillus anthracis*[9] mailings in October of 2001, Congress passed legislation that substantially expanded the scope of the Select Agent Program. The USA PATRIOT Act of 2001 (Public Law 107–56, October 26, 2001) made it an offense for a person to knowingly possess any biological agent, toxin, or delivery system of a type or in a quantity that, under the circumstances, is not reasonably justified by prophylactic, protective, bona fide research, or other peaceful purpose. The Act also established restrictions on the possession or transfer of select agents by "restricted persons," which are individuals with one or more disqualifying factors in their background or behavior (see below).

The provisions of the USA PATRIOT Act were subsequently augmented by the Public Health Security and Bioterrorism Preparedness and Response Act, known as the Bioterrorism Act of 2002 (Public Law 107–188, June 12, 2002). This Act added requirements for regulations governing possession of select agents, including approval for laboratory personnel by the Attorney General following a background check by the Federal Bureau of Investigation (FBI). It also gave the U.S. Department of Agriculture (USDA), through its Animal and Plant Health Inspection Service (APHIS), the authority to regulate the possession, use, and transfer of BSAT materials that relate to plant and animal health and products, complementing the authority granted to CDC for human pathogens. The regulation of select agents and toxins is thus a shared federal responsibility involving HHS/CDC, USDA/APHIS, and the Department of Justice (DOJ). The Bioterrorism Act has been implemented through a series of regulations; the final regulations—42 CFR 73 (human pathogens), 9 CFR 121 (animal pathogens), and 7 CFR 331 (plant pathogens)—became effective in the spring of 2005.[10]

THE CURRENT SELECT AGENT PROGRAM

The current Select Agent Program has a number of components, which are described in the first part of this section. This is followed by a discussion of the operation of the program, such as inspections and inventory requirements.

---

[8]The United States had also weaponized some anti-crop agents, but they were not included in the initial list.

[9]*Bacillus anthracis* is the bacterium that causes anthrax.

[10]Agents that can affect both human and animals, called "overlap agents," are listed in both the CDC and USDA lists.

The List of Select Agents and Toxins

The addition of agents and toxins that could affect animals and plants almost doubled the size of the original select agent list (see Table 2-2 for the current list). In the years since the revised list was issued, the reconstructed 1918 influenza virus was added to the list in 2005; some of the discussion over the wisdom of reconstructing the virus focused on its potential as a biological weapon or bioterror agent. As mentioned in Chapter 1, at the time this report was written, two *Federal Register* notices were seeking comment on proposals

TABLE 2-2 Current List of Select Agents and Toxins

HHS SELECT AGENTS AND TOXINS

Abrin
Botulinum neurotoxins
Botulinum neurotoxin producing species of
    Clostridium
Cercopithecine herpesvirus 1 (Herpes B virus)
Clostridium perfringens epsilon toxin
Coccidioides posadasii/Coccidioides immitis
Conotoxins
Coxiella burnetii
Crimean-Congo haemorrhagic fever virus
Diacetoxyscirpenol
Eastern Equine Encephalitis virus
Ebola virus
Francisella tularensis
Lassa fever virus
Marburg virus
Monkeypox virus
Reconstructed replication competent forms
    of the 1918 pandemic influenza virus
    containing any portion of the coding regions
    of all eight gene segments (Reconstructed
    1918 Influenza virus)
Ricin
Rickettsia prowazekii
Rickettsia rickettsii
Saxitoxin
Shiga-like ribosome inactivating proteins

Shigatoxin
South American Haemorrhagic Fever viruses
    Flexal
    Guanarito
    Junin
    Machupo
    Sabia
Staphylococcal enterotoxins
T-2 toxin
Tetrodotoxin
Tick-borne encephalitis complex (flavi) viruses
    Central European Tick-borne encephalitis
    Far Eastern Tick-borne encephalitis
    Kyasanur Forest disease
    Omsk Hemorrhagic Fever
    Russian Spring and Summer encephalitis
Variola major virus (Smallpox virus)
Variola minor virus (Alastrim)
Yersinia pestis

OVERLAP SELECT AGENTS AND
TOXINS

Bacillus anthracis
Brucella abortus
Brucella melitensis
Brucella suis
Burkholderia mallei (formerly *Pseudomonas
    mallei*)

to add the SARS-associated coronavirus and Chapare virus to the list (HHS 2009ab). The question of adding the SARS and Chapare viruses also illustrates a continuing argument/discussion of whether the list should include agents and toxins that are primarily serious public health threats rather than only those that are likely candidates for use in bioterrorism. The language of the USA PATRIOT Act and the Bioterrorism Preparedness Act speak of agents that pose threats to "public health," but the acts are focused on the threats posed by *bioterrorism* rather than more general infectious diseases. Some of the proposals to stratify the list reflect an effort to focus the Select Agent Program on

---

Burkholderia pseudomallei (formerly *Pseudomonas pseudomallei*)
Hendra virus
Nipah virus
Rift Valley fever virus
Venezuelan Equine Encephalitis virus

USDA SELECT AGENTS AND TOXINS

African horse sickness virus
African swine fever virus
Akabane virus
Avian influenza virus (highly pathogenic)
Bluetongue virus (exotic)
Bovine spongiform encephalopathy agent
Camel pox virus
Classical swine fever virus
Ehrlichia ruminantium (Heartwater)
Foot-and-mouth disease virus
Goat pox virus
Japanese encephalitis virus
Lumpy skin disease virus
Malignant catarrhal fever virus (Alcelaphine herpesvirus type 1)
Menangle virus
Mycoplasma capricolum subspecies *capripneumoniae* (contagious caprine pleuropneumonia)

Mycoplasma mycoides subspecies *mycoides* small colony (*Mmm*SC) (contagious bovine pleuropneumonia)
Peste des petits ruminants virus
Rinderpest virus
Sheep pox virus
Swine vesicular disease virus
Vesicular stomatitis virus (exotic): Indiana subtypes VSV-IN2, VSV-IN3
Virulent Newcastle disease virus[a]

USDA PLANT PROTECTION AND QUARANTINE (PPQ) SELECT AGENTS AND TOXINS

Peronosclerospora philippinensis (*Peronosclerospora sacchari*)
Phoma glycinicola (formerly *Pyrenochaeta glycines*)
Ralstonia solanacearum race 3, biovar 2
Rathayibacter toxicus
Sclerophthora rayssiae var zeae
Synchytrium endobioticum
Xanthomonas oryzae
Xylella fastidiosa (citrus variegated chlorosis strain)

---

[a] A virulent Newcastle disease virus (avian paramyxovirus serotype 1) has an intracerebral pathogenicity index in day-old chicks (*Gallus gallus*) of 0.7 or greater or has an amino acid sequence at the fusion (F) protein cleavage site that is consistent with virulent strains of Newcastle disease virus. A failure to detect a cleavage site that is consistent with virulent strains does not confirm the absence of a virulent virus.
SOURCE: 7 CFR 331, 9 CFR 121, and 42 CFR 73; <http://www.selectagents.gov/>.

those agents and toxins that pose the greatest threats to security (e.g., NSABB 2009). We return to this issue in Chapters 3 and 5 but note it here as part of the context for the operation of the Select Agent Program.

The Security Risk Assessment[11]

A Security Risk Assessment (SRA) is required for all individuals who have access to select agents or toxins, including principal investigators, laboratory staff, and some maintenance personnel. Certain designated officials—the Responsible Official (RO) and the Alternate Responsible Official (ARO), who oversee the program at an individual entity,[12] as well as any owners/controllers of nongovernment entities—must also undergo an SRA even if they will not have personal access to select agents. All registered entities (except for federal, state, or local governmental agencies or accredited public academic institutions) must also undergo an SRA.[13]

The SRA for individuals is carried out by the FBI's Criminal Justice Information Services (CJIS) Division. An individual's SRA is valid for five years unless terminated sooner by CDC, APHIS, or the employer. The SRA is tied to the entity for which the individual works; it cannot be transferred if she or he moves to another BSAT facility. A certificate of registration for an entity is valid for a maximum of three years. The entity, the RO, ARO, and individuals that own or control the entity must obtain security risk assessment approval each time the certificate of registration is renewed.

The purpose of the SRA is to determine whether an applicant has any of the factors that would prohibit him or her from working with select agents,

---

[11]Unless otherwise noted, this section is based on a briefing prepared for the committee by Robbin Weyant, Division of Select Agents and Toxins, Coordinating Office for Terrorism Preparedness and Emergency Response, Centers for Disease Control and Prevention, and Elizabeth Snyder, Criminal Justice Information Services, Federal Bureau of Investigation, and presented on June 29, 2009 (Weyant and Snyder 2009).

[12]"An entity is defined as any government agency (federal, state, or local), academic institution, corporation, company, partnership, society, association, firm, sole proprietorship, or other legal entity. An entity is thus not limited to a single facility or to a single laboratory. An entity may possess one or multiple facilities, each facility containing one or multiple laboratories" (Gottron and Shea 2009:2). Clinical laboratories were granted a special exemption to permit them to legally isolate and identify (and thereby possess) select agents cultured from patients as part of the medical diagnostic process, even if they were not registered to possess select agents, and to handle those materials that were part of proficiency training. This was considered essential for medical diagnoses where there is no way to predict what disease a patient might have, thereby precluding the ability to register for specific select agents. The clinical laboratories, however, are mandated to destroy any select agents or transfer them to a registered laboratory that is permitted to possess them within specified periods, and they must also notify public health authorities whenever a select agent has been isolated and identified (Deminn 2007:547).

[13]The government facilities are presumed to have equivalent security procedures in place.

based upon the exclusions enumerated in relevant legislation. An individual is considered a "restricted person" under the USA PATRIOT Act if he or she:[14]

- Is under indictment for a crime punishable by imprisonment for a term exceeding one year or has been convicted in any court of a crime punishable by imprisonment for a term exceeding one year.
- Has received a dishonorable discharge from the U.S. military. This provision ensures that those who commit comparable crimes while in the military will also be denied access to BSAT materials.
- Is a fugitive from justice.
- Is an unlawful user of any controlled substance (as defined in section 102 of the Controlled Substances Act (21 USC 802)).
- Has been adjudicated as a mental defective or has been committed to any mental institution. The prohibition is based on specific legal distinctions that make this a small category of individuals.[15]
- Is an alien illegally or unlawfully in the United States.
- Is an alien (other than an alien lawfully admitted for permanent residence) who is a national of a country that has repeatedly provided support for acts of international terrorism. This is operationalized as nationals of countries formally designated as state sponsors of terrorism. Currently there are four such countries: Cuba, Iran, Sudan, and Syria (Department of State 2009b).[16]

Unlike formal security clearances, foreign nationals are thus eligible for access to BSAT.[17]

Additionally, under the Bioterrorism Preparedness Act, an individual could

---

[14]According to several accounts of the creation of the USA PATRIOT Act, including those at the public consultation for the Executive Order Working Group on Strengthening the Biosecurity of the United States, the list of disqualifying factors was based on those in the Brady Handgun Violence Prevention Act (Public Law 103–159, November 30, 1993), on the principle that anyone who would not be allowed to own a handgun should not be permitted access to dangerous pathogens.

[15]"Adjudicated as a mental defective. (a) A determination by a court, board, commission, or other lawful authority that a person, as a result of marked subnormal intelligence, or mental illness, incompetency, condition, or disease: (1) Is a danger to himself or to others; or (2) Lacks the mental capacity to contract or manage his own affairs. (b) The term shall include (1) A finding of insanity by a court in a criminal case; and (2) Those persons found incompetent to stand trial or found not guilty by reason of lack of mental responsibility pursuant to articles 50a and 72b of the Uniform Code of Military Justice, 10 USC 850a, 876b" (27 CFR 478.11). See Chapter 4 for further discussion.

[16]At the time the SRA was developed, there were seven countries on the list, the four current ones and Iraq, Libya, and North Korea.

[17]Foreign nationals who possess unique skills or knowledge can obtain special security clearances that provide limited access for specific purposes.

not have access to select agents if he or she is "reasonably suspected" by any federal law enforcement or intelligence agency of:

- Committing a crime specified in 18 USC 2332b(g)(5);[18]
- Having a knowing involvement with an organization that engages in domestic or international terrorism (as defined in 18 USC 2331) or with any other organization that engages in intentional crimes of violence; or
- Being an agent of a foreign power (as defined in 50 USC 1801).[19]

Evidence of any of the disqualifying factors leads to a permanent denial of access to select agents and toxins, without statute of limitations or sunset provision on any of the prohibitions. An individual whose request for access is denied receives a formal, written notification, which will contain the specific reason for the denial. An individual may appeal the denial in writing to CDC or APHIS within 30 calendar days if he or she believes the information used as the basis for the denial is incorrect (such as in the case of mistaken identification). The appeal must state the factual basis for the request to overturn the denial and provide supporting documentation. If the denial was based on material produced by the CJIS review of federal databases, the appeal will be forwarded to the FBI.

According to data provided to the committee by CDC, USDA, and the FBI, as of September 10, 2009, a total 31,349 individual applications had been processed by CDC and USDA since the beginning of the program. Of these, 1,860 were withdrawn prior to renewal and 192 applicants were denied as restricted persons. Fifty-eight restricted individuals had appealed the denials, of which 36 denials were sustained and 22 were overturned. Of those denied access, 158 individuals were restricted for a single prohibited factor, while 12 had multiple disqualifying factors. In just under 70 percent of the cases, the disqualifying factor was being convicted of a crime with greater than a one-year imprisonment as the potential penalty.

According to information provided by FBI/CJIS to the committee, as of September 10, 2009, approximately 13,609 individuals had *active* SRA approval for access to select agents regulated by CDC or APHIS. At present, the average turnaround time for an SRA screening is 31 days, down from an average of 61 days in 2008 (Weyant 2009). CDC did not keep records on processing times from the start of the program, but, in the initial phases of implementation, the processing time was much longer, sometimes months, because the

---

[18]This section defines "Federal crimes of terrorism."

[19]This section refers to persons who engage in clandestine intelligence activities on behalf of a foreign government or in support of a foreign group that engages in international terrorism or preparations for it.

need to review so many people initially overwhelmed the system. Some of the committee's discussions and site visits suggested that the process still can take several months, although now only for specific cases.

Implementing the SRA

The assessment of whether an individual has any of these disqualifying factors is based on responses to questions on the SRA application (FBI Form FD-961; see Appendix D) along with a fingerprint check and a search of a wide range of federal databases to identify disqualifying background/activities.[20] Box 2-1 contains a list of the databases searched as part of the SRA.

SRA screening to identify those with criminal records relies on the standard criminal databases maintained by the FBI and Department of Homeland Security (DHS) for these purposes and used to conduct routine suitability or security screening for other federal agencies. These databases are also available to, and widely used by, employers outside the federal government. It is noted in a DOJ study of the government's criminal history databases that the FBI handles more fingerprint checks for noncriminal justice purposes than for those related to law enforcement (DOJ 2006:3); for example, it processed approximately 10 million such requests in 2005.

There is widespread interest in obtaining access to criminal history record information from reliable sources for the purpose of screening an individual's suitability for employment, licensing, or placement in positions of trust. The interest comes from private and public employers, as well as non-profit organizations that place employees and volunteers to work with vulnerable populations such as children, the elderly, and disabled persons. The interest is based on a desire or perceived need to evaluate the risk of hiring or placing someone with a criminal record in particular positions and is intended to protect employees, customers, vulnerable persons, and business assets. Employers and organizations are subject to potential liability under negligent hiring doctrines if they fail to exercise due diligence in determining whether an applicant has a criminal history that is relevant to the responsibilities of a job and determining whether placement of the individual in the position would create an unreasonable risk to other employees or the public. In addition to addressing this litigation risk, employers want to assess the risks to their assets and reputations posed by placing persons with criminal histories in certain positions. To meet these business needs, employers can and frequently

---

[20]One of the questions on the application is "Have you ever been adjudicated as a mental defective or been committed to any mental institution?" For anyone who answers "yes," "a complete copy of medical records regarding the commitment will be required." This is designed to address the difficulties of obtaining such records, which are largely held by states and may be subject to significant privacy restrictions under state or federal law.

BOX 2-1
Databases Used to Carry Out SRA Investigations

National Crime Information Center Files
- Foreign Fugitive File
- Deported Felon File
- Protection Order File
- Wanted Person File
- U.S. Secret Service Protective File
- SENTRY File (Bureau of Prisons)
- Convicted Person on Supervised Release File
- Convicted Sexual Offender Registry
- Violent Gang and Terrorist Organizations File

Interstate Identification Index (III; "Triple I"): state/local criminal history

Foreign Terrorist Tracking Task Force
- Terrorist Screening Center Database (TSDB)
- Transportation Security Administration's No Fly and Selectee databases

Automated Case Support (ACS): FBI case file database [Extra investigative effort is put into instances when the ACS contains significant derogatory information on the individual]

Bureau of Immigration and Customs Enforcement's Law Enforcement Support Center databases (Foreign-born candidates):
- Central Index System (CIS)
- Computer Linked Application Information Management System (CLAIMS)
- Deportable Alien Control System (DACS)
- National Automated Immigration Lookout System (NAILS II)
- Nonimmigrant Information System (NIIS)
- Student and Exchange Visitor Information System (SEVIS)
- Redesigned Naturalization Application Casework System (RNACS)
- Refugee, Asylum, and Parole System (RAPS)
- Enforcement Case Tracking System (ENFORCE)
- Treasury Enforcement Communications System (TECS)

SOURCE: Weyant and Snyder 2009.

---

do ask applicants whether they have a criminal history. Such employers and organizations want access to criminal history records to determine whether applicants are answering the question about their criminal history truthfully and completely. They believe that having access to good sources of criminal history information is the only way the interest in performing due diligence to protect employees, assets, and the public can be served. Public employers' need for the information often goes beyond considering job suitability and includes security clearance determinations. There also has been a growing use of criminal history screening in certain sectors of the economy related to counterterrorism efforts. (DOJ 2006:1)

It should be noted that questions have been raised about the accuracy and completeness of the standard databases. The same DOJ study cited above found, for example, that although the Interstate Identification Index (III or "Triple I") system—one of those used in the SRA screening—was "quite comprehensive in its coverage of nationwide *arrest* records for serious offenses," it was missing information about the *final disposition* of about 50 percent of the cases in the database (DOJ 2006:3; emphasis added).[21] In addition, a recent study of the Transportation Security Agency's (TSA) screening of port workers found substantial delays in the processing time and some evidence of a problem with incorrectly identifying some candidates as having criminal records, although the appeal process seemed able to address most of the mistakes (NELP 2009).

The SRA relies upon 10 databases maintained by the Bureau of Immigration and Customs Enforcement's Law Enforcement Support Center to identify whether non-U.S. citizens are in the country legally. These are the standard databases used throughout the federal government for screening an individual's immigration status. Increased security concerns after the September 11, 2001, terrorist attacks and ongoing debate over immigration policy have led to significant investment in identifying and tracking foreign nationals visiting or living in the United States. This has resulted in substantial improvements in the databases, although concerns about errors and misidentifications remain, for example because the same name may be shared by multiple individuals or because of confusion caused by the transliterated spelling of names. It should be noted that the policy discussions related to screening non-U.S. citizens have focused largely on the difficulty of acquiring information about an individual's life prior to his or her arrival in the United States rather than the adequacy of the databases for tracking people once they have arrived. The report on personnel reliability from the National Science Advisory Board for Biosecurity (NSABB), for example, recommends more rigorous screening for foreign nationals but does not make specific suggestions for how that should be implemented (NSABB 2009).

As mentioned above, the Bioterrorism Preparedness Act added prohibitions for work with BSAT materials that concern involvement in crimes of terrorism or with groups who commit acts of violence. Additional databases, such as the Violent Gang and Terrorist Organizations File, the Terrorist Screening

---

[21]The Triple I contains arrest records for serious offenses from all states and territories, as well as from federal and international criminal justice agencies.

Center Database, and TSA's No Fly and Selectee databases, are used for this part of the screening.

The issue of illegal drug use is captured by SRA screening in two ways. Criminal database searches would identify anyone convicted of a drug-related felony offense as part of the general criminal check.[22] The application also asks, "Are you an unlawful user of any controlled substance (as defined in Section 102 of the Controlled Substance Act [21 U.S.C. 802])?" and anyone answering affirmatively is disqualified. There is no attempt to verify the accuracy of statements about current use, however.

In addition to the initial review and a new review every five years or if an individual moves to a new entity, "the FBI is automatically notified when an individual with a favorable SRA is arrested and fingerprinted or checked against criminal databases for whatever reason. The FBI also monitors individuals with favorable SRAs for criminal activity or terrorist ties by periodically cross-checking their names and fingerprints against federal databases. Access to select agents can be denied, limited, or revoked at any time by the institutional RO or ARO, CDC, or USDA if deemed appropriate" (NSABB 2009:3).

Recognizing the importance of collaboration in scientific research, procedures are in place to enable an individual with a current SRA to visit another registered entity. The RO of the home entity must notify the RO of the receiving entity in writing that the proposed visitor has an active approved SRA. The RO of the entity that will be hosting the visitor must submit this letter and a request to amend the registration to the lead agency (APHIS or CDC), which will approve or deny the amendment. Once the visit is complete, the receiving entity RO should amend the entity's registration to remove the visitor.[23]

Participation in the Program[24]

The Select Agent Program requires "registration of facilities including government agencies, universities, research institutions, and commercial entities

---

[22]"Of current federal prisoners, 55 percent are serving time for drug offenses" (Washington Post 2009).

[23]By comparison, in most cases the transfer of security clearances can be done between the home and host institution without formal permission of the agency that issued the clearance, in part because security officers at these institutions have access to secure databases to verify the clearances.

[24]Under current regulations, entities that do not at any time have more than the following aggregate amounts of a toxin (in the purified form or in combinations of pure and impure forms) under the control of a principal investigator are excluded from requirements of the regulation: abrin (100 mg), botulinum neurotoxin (0.5 mg), *Clostridium perfringens* epsilon toxin (100 mg), conotoxins (100 mg), diacetoxyscirpenol (1,000 mg), ricin (100 mg), saxitoxin (100 mg), shiga-like ribosome inactivating proteins (100 mg), shigatoxin (100 mg), staphylococcal enterotoxin (5 mg), tetrodotoxin (100 mg), and T-2 (1,000 mg) <http://www.selectagents.gov/Permissible%20Toxin%20Amounts.html>.

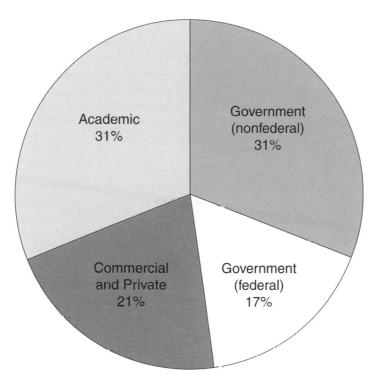

FIGURE 2-1 Entities registered to work with select agents and toxins. Data provided by APHIS and CDC.

that possess, use or transfer biological agents and toxins that pose a significant threat to public, animal or plant health, or to animal or plant products."[25] Material provided to the committee by CDC and APHIS showed that, as of early September 2009, 388 entities had received authorization to work with select agents and toxins (see Figure 2-1), of which 84 percent were registered with CDC and 16 percent with APHIS (Kielbauch et al. 2009). The largest single components were nonfederal government laboratories (121 entities) and academic entities (120 entities). Federal laboratories comprised 65 entities, and the remaining component was commercial and private entities (82 entities) (Kielbauch et al. 2009). There is no directory of the facilities within these categories, because the names of the registered facilities are not made public.

One of the most serious challenges to implementing the Select Agent Program is the sheer diversity of facilities that work with select agents. This diversity extends both within and across the four main categories of entities

[25]<http://www.selectagents.gov/>.

and includes, among other factors, size, the type and mix of personnel, the characteristics of work carried out, the variety of sponsors for the work, and the variety of BSAT materials each holds. What follows is a brief description to illustrate the range of facilities encompassed within the program. Given its charge, the committee's interest focused on those facilities that have the conduct of research as their primary focus, which includes at least some entities within each category.

Academic Entities

The public consultations organized to inform related reports, public discussions held by the committee, and the committee's site visits revealed significant diversity among academic laboratories. In addition to research facilities at universities and academic medical centers, "academic entities" include a few nonprofit research facilities such as the Southwest Foundation for Biomedical Research in San Antonio, Texas.

Three of the five currently operating BSL-4 laboratories fall within this category: the Southwest Foundation, Georgia State University, and the University of Texas Medical Branch at Galveston (UTMB). UTMB has also constructed a new, larger BSL-4 laboratory—the Galveston National Laboratory—that is awaiting final certification, and Boston University has completed construction of a BSL-4 laboratory but is awaiting resolution of legal issues surrounding an environmental impact statement before it can be opened (NRC 2007a).

Most select agent laboratories involve research at the BSL-3 safety level. In most cases, select agent laboratories comprise a small fraction of the total number of biological laboratories on a campus; for example, the biosafety officer from Vanderbilt University informed the committee that there is one select agent laboratory out of a total of 500 at her university (Burnett 2009).[26] The size, staffing, and type of research conducted by academic select agent laboratories varies substantially. The NIH-funded New England Regional Center of Excellence National Screening Laboratory, located at Harvard Medical School, reported an SRA-cleared staff of 12, including four technical staff members, four animal care staff, two administrators, and two postdoctoral researchers. George Mason University anticipates registering 30-40 individuals employed by the university, as well as several contracted animal care technicians when its Biomedical Research Laboratory is operational in 2010.

Federal Entities

The federal government operates the largest category of entities that are part of the Select Agent Program. NIH and CDC within HHS, USDA, the

---

[26]Most academic research laboratories that conduct non-select agent research are BSL-1, BSL-2, and BSL-3 facilities.

Department of Defense (DOD) and the separate military services, the Department of Energy (DOE), and the Environmental Protection Agency—operate laboratories conducting research on select agents and toxins. Some of the facilities conduct specifically defense-related research, such as DOD's Edgewood Chemical Biological Center, which focuses on defensive needs of the warfighter, including development of protective equipment. Others, such as DHS' National Biodefense Analysis and Countermeasures Center (NBACC), which will focus on threat characterization and bioforensics research when it is operational, reflect increasing concern with homeland security. And others, such as NIH's Rocky Mountain Laboratories or USDA's National Plant Germplasm and Biotechnology Laboratory, primarily carry out research related to broader threats of select agents and toxins to human or animal health or plant and animal products.

Private and Commercial Entities

Private nonprofit and commercial for-profit entities include pharmaceutical and biotechnology firms carrying out research with select agents and toxins as part of developing a range of diagnostics, vaccines, and related therapeutics. Some of this research is supported by the federal government, and some is part of product development by companies for the private market. A number of the entities in this category also specialize in carrying out contracted research for multiple sponsors, pubic and private, who do not choose to maintain a laboratory. It is these facilities that are particularly affected by multiple inspections and conflicting security requirements imposed by different agencies.

State and Local Government Entities

State and local government laboratories are generally quite different from the other entities in the program. Most are state and local public health laboratories, part of the Laboratory Response Network (LRN), a national network of approximately 150 laboratories created in 1999 by HHS to respond to chemical and biological terrorism and other public health emergencies.[27] Although some of the larger state laboratories conduct research, most of these laboratories do not conduct regular work with select agents. They are included in the Select Agent Program because they might encounter select agents as part of routine diagnostic work and because they maintain reference strain collections to confirm specific select agents and toxins as part of the LRN.

---

[27]The LRN also includes federal, military, environmental, veterinary, and food-testing laboratories. See <http://www.bt.cdc.gov/lrn/factsheet.asp> for more information.

Reference labs, sometimes referred to as "confirmatory reference," can perform tests to detect and confirm the presence of a threat agent. These labs ensure a timely local response in the event of a terrorist incident. Rather than having to rely on confirmation from labs at CDC, reference labs are capable of producing conclusive results. This allows local authorities to respond quickly to emergencies. (CDC 2009)

Operation of the Program[28]

All entities registered to possess select agents and toxins must develop a security plan to safeguard the select agents in their possession against unauthorized access, theft, or loss. There must also be a biosafety plan, commensurate with risks posed by the agents or toxins, to safeguard against their release in the laboratory or to the wider environment. The existing BMBL, *NIH Guidelines for Research Involving Recombinant DNA Molecules*, and regulations from the Occupational Safety and Health Administration provide guidance for developing the safety plan (current biosafety guidelines were discussed in more detail earlier). Laboratories registered to work with toxins must also have a chemical hygiene plan. Finally, an incident response plan is required to be sure the entity is prepared to deal with the consequences of an accident or deliberate act.

Inspections

"An important tenet of the CDC Select Agent Program is that it treats all registered entities the same—whether that lab is a commercial lab, state or local public health lab, or a federal lab (including CDC and Department of Defense labs)" (Besser 2007); the same principle applies for those entities supervised by APHIS for USDA. Implementation of this principle means that common standards are employed, the checklists used during inspections and the requirements laboratories are expected to meet are the same, and there is a consistent determination of what will trigger a referral of any laboratory to the HHS Office of Inspector General (OIG) or APHIS Investigative and Enforcement Services (APHIS-IES) for possible violations of the regulations. CDC and APHIS provide training and compliance assistance to help individuals and entities understand and meet the requirements of the program.

Inspections are the primary means by which compliance with the regulations is confirmed. Routine inspections are carried out every three years, with additional inspections taking place whenever an entity desires to make a significant change to its select agent registration, such as changes in currently registered laboratories or additional new laboratories that require registration.

---

[28]This section is drawn in large part from Congressional testimony given by CDC official Richard E. Besser (2007).

Other inspections may take place as follow-up resulting from audits by federal partners or investigation of potential biosafety or security concerns that could affect public health and safety. Between 2003, when the interim regulations for the program were released, and early summer 2009, CDC conducted 840 select agent inspections and APHIS 324 inspections. The inspections were frequently done in collaboration with each other and other federal agencies.

The procedure for routine inspections involves an extensive review of laboratory safety and security as related to the possession, use, and transfer of select agents, using checklists based on the select agent regulations and recognized safety standards.[29] Inspectors observe the physical safety and security components of the facility, examine the documentation available, and interview laboratory personnel to collect information used to complete the checklists. Results of the CDC or APHIS inspection are provided to the institution in a written report, and entities must respond within a specified time to any deficiencies noted in the inspection report, with documentation of how they have resolved those deficiencies. If the deficiencies are considered serious enough, a verification site visit would be used to confirm that the problems have been corrected.

Several types of enforcement actions can occur in cases where there are possible violations of the select agent regulations:

- *Administrative actions:* a registered entity's certificate of registration can be suspended or revoked (a suspension can include work at a registered entity or be specific to particular agents or particular types of experiments). An entity's application to possess, use, or transfer select agents can also be denied.
- *Civil or criminal penalties:* Civil monetary penalties (up to $250,000 for an individual for each violation and up to $500,000 for an entity for each violation) or criminal enforcement actions (imprisonment for up to five years, a fine, or both) are possible for more serious violations.
- *Referral to FBI:* Possible violations involving criminal negligence or a suspicious activity or person are referred to the FBI for further investigation.[30]

The Select Agent Program also promotes laboratory safety and security by providing technical assistance and guidance to registered entities, which in-

---

[29]The checklists can be found at <http://www.selectagents.gov/>.

[30]As of December 19, 2008, APHIS had referred 36 entities to APHIS-IES for violation of the select agent regulations. APHIS-IES had levied $109,250 in civil monetary penalties against seven of the entities. As of early September 2009, the HHS OIG had reported a total of $1,997,000 in fines levied against 13 organizations for failure to comply with various aspects of the select agent regulations (see <http://oig.hhs.gov/fraud/enforcement/cmp/agents_toxins.asp> for descriptions of the cases). HHS and USDA have not referred any violations of the select agent regulations to DOJ for criminal prosecution.

cludes presenting workshops, having a primary point of contact assigned to each entity, developing frequently asked questions that are posted on the program Web site, and making technical presentations at meetings and conferences.

Inventory

Entities that possess BSAT materials are required to keep specific kinds of records and other information about the materials in their possession. (Addtional information about current policies is included in Chapter 5.) An accurate, current inventory is required for each agent in long-term storage, that is, maintained in a condition that keeps them viable for future use.[31] Entities are also required to have protocols in place for transfer and accountability of inventories when the investigator responsible for the inventory leaves, including change in employment, retirement, death, sabbatical, or other reasons, and no longer has an active role in the entity.

Reports of Theft, Loss, and Release

All reports of theft, loss, or release of select agents are investigated to ensure that public health and safety are protected. From 2003 until the end of September 2009, there were 154 incidents reported to CDC and USDA through the Select Agent Program's theft, loss, and release reporting system. Follow-up investigations conducted by HHS, USDA, and the FBI determined that there were no confirmed losses or theft of a select agent. There were three confirmed releases of a select agent, and these were all identified by illnesses in five laboratory workers as a result of working with the agents (Besser 2007).

OTHER FEDERAL REGULATIONS RELATED TO BSAT RESEARCH

The Select Agent Program's regulations apply to all federal agencies and departments. Most of the federal agencies conducting or supporting BSAT research—including DOD and the military services, NIH, CDC, DOE, and USDA—currently have additional security-related policies or regulations in place beyond these requirements. At least some of the agencies require that the entities conducting BSAT research they fund via contracts and grants also apply these practices. This section contains a brief summary of these additional polices and regulations, based largely on information provided by the Executive Order (EO) Working Group on Strengthening the Biosecurity of the United

---

[31]A definition of "long-term storage" can be found at <http://www.selectagents.gov/LongTermStorage.html>. See Box 5-2 for required elements in inventory records.

States. After a brief recap of personnel reliability and physical security, the section is organized by agency.[32]

Physical Security and Personnel Reliability

Physical Security

Physical security programs are intended to prevent unauthorized access to BSAT materials. They are largely but not exclusively addressed to combating the outsider threat. Three common elements of physical security programs are: (a) access controls, which include security for the perimeter, points of entry, and the interior of the facility; (b) information systems control, which includes information technology, protection of infrastructure and hardware, assuring reliability of information technology personnel or vendors, and inventory protection; and (c) inventory controls, which includes managing and tracking both the BSAT materials and the data about them.

Personnel Reliability

A "personnel reliability program" (PRP) is the general term used to describe policies intended to ensure that individuals who are given access to BSAT materials are worthy of that trust. There may be many qualities that define a "reliable" employee; the NSABB, for example, concluded that trustworthy, responsible employees would:

- Be free of felony convictions;
- Have no domestic or international terrorist ties;
- Have no history of scientific or professional misconduct in the workplace;
- Possess emotional stability and capacity for sound judgment;
- Have a positive attitude toward safety and security measures, and standard operating procedures; and
- Be free of vulnerability to coercion (NSABB 2009).[33]

---

[32] The section does not address issues related to transportation, which the committee has decided not to include in its report.

[33] The report goes on to say that: "The NSABB considers these to be reasonable characteristics for individuals with access to select agents and toxins. It found, however, that some of the characteristics were exceedingly difficult to measure in any objective way and that it was unclear whether these characteristics were suitable surrogates (or predictors) for not posing an insider threat. Furthermore, as it considered the potential utility of the various assessments commonly utilized in PRPs, it found little evidence to suggest that personnel reliability assessments going beyond the SRA and other institutional background checks that are already in place would correlate with, or effectively identify, an insider threat. In addition, as was the case with the optimal personnel characteristics, there were no objective criteria for translating the information gathered from a given assessment into a determination of reliability" (NSABB 2009:8).

A PRP may apply in the pre-employment screening phase as part of the hiring process or to measures that apply while an individual is working or both. At hiring, applicants may be screened for drug or alcohol (ab)use; undergo medical examinations, psychological evaluations, credit checks, and reviews of past employment or service records; be required to take psychological tests or polygraph exams; and undergo background investigations, possibly including a formal security clearance process. Drug and alcohol screening might continue during employment, along with a range of mechanisms for continuous monitoring, including self-reporting and peer reporting and periodic updates of the checks and assessments done prior to hiring. A number of the personnel reliability programs carried out by federal agencies require various types of formal background investigations; these are described briefly in Box 2-2.

The federal Office of Personnel Management (OPM) currently conducts 90 percent of the suitability and security clearance investigations for 100 federal agencies, using approximately 9,000 investigators to conduct 2.2 million background investigations each year (Crowley 2009). Of these, approximately 750,000 are national security clearance investigations (GAO 2009b:1). The process for determining access to classified information rests on a series of executive orders dating back to the 1950s; the current process is based on EO 12968, issued in August 1995.[34]

Examples of Federal Agency Physical Security and Personnel Reliability Requirements

This section is intended to provide a sampling, agency by agency, of some of the additional provisions that federal agencies have put in place to increase security for BSAT materials beyond the requirements of the Select Agent Program. Some of the measures simply reflect the agency's practice for many of its employees—such as requiring security clearances—and are not limited to those working with BSAT materials. Others, such as the Army's Biological Surety Program, are specifically directed at BSAT research.

U.S. Department of Agriculture

USDA's PRP policy covers personnel who work with BSAT at the BSL-3 level, with the extent of the background investigation determined by the level of risk. Continuous evaluation by managers and required self-reporting are part

---

[34]In June 2008, President George W. Bush issued EO 13467, which was intended to "ensure an efficient, practical, reciprocal, and aligned system for investigating and determining suitability for Government employment, contractor employee fitness, and eligibility for access to classified information" (White House 2008). This EO set in motion a number of reforms, but these are beyond the scope of this study.

BOX 2-2
Types of Personnel Security Investigations

National Agency Check. The National Agency Check (NAC) consists of searches of the Security/Sustainability Investigation Index and the Defense Clearance and Investigations Index, as well as the FBI Identification Division's name and fingerprint files, and other files as necessary. These are conducted by the Office of Personnel Management.

National Agency Check and Inquiries. The National Agency Check and Inquiries (NACI) is a basic investigation required for all new federal employees. It consists of the National Agency Check investigation, as well as written inquiries and record searches covering specific areas of a person's background during the past five years. Inquiries are sent to employers, schools attended, references given, and local law enforcement authorities.

NACI and Credit. The NACI and Credit (NACIC) consists of the NACI with the addition of a credit record check.

Access NACI. The Access NACI (ANACI) consists of a required initial investigation for federal employees who will need access to classified national security information at the Confidential and Secret levels. The ANACI includes the NACIC with additional law enforcement agency checks.

NAC with Local Agency Check and Credit. The NAC with Local Agency Check and Credit (NACLC) is the initial investigation for government contractors at the Confidential and Secret national security access levels. The NACLC is also used to meet reinvestigation requirements for all individuals holding Confidential and Secret clearances.

Single Scope Background Investigation. The Single Scope Background Investigation (SSBI) is a government-wide investigation required for all personnel needing access to Top Secret classified national security information. The SSBI covers the last seven years of the person's activities and includes verification of citizenship and the date and place of birth. In addition, the SSBI conducts national agency records checks on the person's spouse or cohabitant and interviews with selected references and former spouses.

SSBI-Periodic Reinvestigation. The SSBI-Periodic Reinvestigation (SSBI-PR) is required every five years for personnel with Top Secret security clearances.

Schedule. Investigations are nominally conducted on a five-year reinvestigation schedule. In some cases, a specific type of national security clearance may call for a reinvestigation on a faster schedule. Investigations for collateral Secret and lower clearances sometimes exceed five years due to budgeting or workload constraints.

SOURCE: Appendix B in DSB 2009.

of the system. USDA employees have an employee assistance program in place to provide counseling services.

Agricultural Research Service (ARS)   ARS policies regarding BSAT inventories are included in a USDA manual that sets out procedures for BSL-3 laboratories (USDA 2002). Three types of records are required for facilities to demonstrate proper accountability for BSAT materials. The first is the National Pathogen Inventory, a system that is intended to enable managers to determine quickly what pathogens are in use at their facility. The second requirement is for a detailed inventory of records, both current and historical, which must be retained for five years. Facilities are also expected to have a materials accountability program for experimental and working samples they maintain.

Department of Defense

DOD has a number of special instructions related to physical security, some across the military and others specific to individual services. The DOD personnel reliability program includes one kind of security clearance and background investigation for military personnel and contractors, the National Agency Check with Local Agency Check and Credit Check (NACLC), to confirm information supplied by applicants, in most cases going back seven years. DOD also has another, comparable process for federal civilian employees, the Access National Agency Check and Inquiries (ANACI). Reinvestigations are done every five years if a clearance is to be continued. Unlike the nuclear and chemical PRP programs, foreign nationals are allowed access in the biological PRP. Peer and self-reporting of any potentially disqualifying information are required.

As the Defense Science Board (DSB) report on the DOD biological safety and security program notes, DOD's nuclear surety program has two categories for personnel in its workforce and each category is subject to different background investigations. "A *critical* position is someone who possesses both technical knowledge and access to the nuclear weapon/system (e.g., launch officers, maintenance personnel, etc.) and *non-critical* is an individual who possesses access but not technical knowledge (e.g., guard forces)" (DSB 2009:31; emphasis added).

Department of the Army   The Army applies additional physical security measures to all of the BSAT laboratories and facilities it owns or controls. The same rules apply to the major Army commands, and to contractors who have received BSAT materials from DOD; the latter requirement caused some questions and concerns during the public consultations. Commenters suggested a formal vulnerability assessment for both outsider and insider threats, based on Army and DOD guidance. Rather than the site-specific plan required by the Select

Agent Program, an extensive list of specific features, subject to annual review, is required for both the facility and the rooms and laboratories where BSAT work takes place under Army regulations. Two people are required to access reference stocks, for example, but there is no current requirement for cameras.

In the summer of 2008, the U.S. Army issued a new set of regulations governing personnel reliability: AR 50-1, the "Biological Surety" program (Department of the Army 2008). In addition to the security clearances required for those granted access to BSAT, AR 50-1 gives responsible officials substantial new discretion to deny or remove access. Some of the provisions, which describe factors well beyond those in the current SRA that would disqualify an individual access to BSAT, are potentially exceedingly expansive. For example, one can be disqualified for an "inappropriate attitude, conduct, or behavior" (Department of the Army 2008:10).

Department of Energy

Of the five DOE national laboratories that work with BSAT materials, all require elements of a PRP beyond what is required by the Select Agent Program. Polygraph examinations are not required, in contrast to DOE's nuclear activities where individuals with access to certain kinds of highly classified information may be required to undergo polygraph testing. Different laboratories use one of two different programs, both of which include an annual medical exam and psychological evaluation, annual credit and criminal records checks, and various training components.

As part of its Worker Safety and Health Program outlined in 10 CFR 851, DOE sets out requirements for its contractors to maintain an inventory and to submit an annual report on its status to the contractor's Institutional Biosafety Committee, which is also provided to the relevant DOE field and area offices. Copies of reports of transfers of BSAT materials, including notification when the transfer is complete, are also sent to the relevant DOE field office.

Department of Health and Human Services

National Institutes of Health The NIH Biological Surety Program covers all personnel who work in BSL-4 facilities and anyone who works in certain other designated facilities. These individuals must undergo a Collective Foreign Threats Assessment, which searches most of the databases used in the SRA and several others. The level of additional background investigation conducted is based on the sensitivity of the job responsibilities. For those working in designated facilities including BSL-4 laboratories, NIH may require a two-person rule or buddy system, although this is for occupational health and safety rather than security. On the job, continuous monitoring, self- and peer-reporting, training, and medical and behavioral health exams are required, although again the primary motive is safety.

Centers for Disease Control and Prevention   CDC is currently developing a personnel screening program and monitoring program for all employees who work with or have access to BSAT that is expected to include an ANACI security clearance process, drug testing and occupational health screening, and self and peer-reporting.

Smallpox was declared eradicated in 1980, and in 1983, two centers—CDC and one in Russia—were authorized by the World Health Assembly as the sole entities able to house or conduct research on smallpox. The original plan was to destroy all remaining stocks after 10 years, but this was changed to create a standing World Health Organization (WHO) Advisory Committee on Variola Virus Research that monitors the state of all research, grants permission to conduct specific experiments, and reports to the World Health Assembly annually (IOM 2009:1-2; WHO 2008).

Department of Homeland Security

Although it does not have a formal PRP program, DHS already requires all employees to have a minimum of a Secret security clearance. The two DHS laboratories that work with BSAT materials—the Plum Island Animal Disease Center and NBACC—require drug screening for all potential employees. NBACC also requires a reliability screening by a senior laboratory manager, which includes personnel and medical information. Employees are required to report potentially disqualifying information.

REGULATIONS AND PRACTICES OUTSIDE THE UNITED STATES

Life sciences research is an increasingly global enterprise, including that involving BSAT materials (NRC 2006). A number of regional and international organizations have developed standards and practices that are relevant to the conduct of BSAT research. In addition, a number of countries have regulations and guidance that either govern work with BSAT materials directly or have an impact on how such work is conducted. What happens outside the United States needs to be considered because it may provide useful ideas or models. Moreover, international collaboration benefits from harmonized standards and practices.

The Geneva Protocol, the BWC, and UN Security Council Resolutions 1540 and 1810 were described earlier in this chapter. In addition, a number of formal and informal groups address parts of BSAT research.

International and Regional Activities

The Australia Group

The Australia Group (AG) is an informal forum of 40 countries and the European Commission that "through the harmonisation of export controls, seeks to ensure that exports do not contribute to the development of chemical or biological weapons."[35] The AG was formed in 1985 in response to a proposal from Australia to improve consultation over export controls on chemical weapons precursors after the use of chemical weapons in the Iran-Iraq War. Biological materials and equipment were included in the AG's concerns in the early 1990s. The AG maintains "common control lists" for dual use biological equipment and related technology and software, biological agents, and plant and animal pathogens to provide the basis for encouraging standard national export licensing regulations.

European Committee for Standardization

The European Committee for Standardization/Comité Européen de Normalisation (CEN) is a private, nonprofit organization that seeks to promote the development of standards in order to reduce trade barriers, promote safety, allow interoperability of products, systems and services, and promote common technical understanding. "All standards help build the 'soft infrastructure' of modern, innovative economies. They provide certainty, references, and benchmarks for designers, engineers and service providers. They give 'an optimum degree of order.'"[36] Much of CEN's effort is carried out through workshops that reach consensus on voluntary standards (called CEN Workshop Agreements) that can be applied internationally and do not have the force of regulation. In 2008, CEN published its *International Laboratory Biorisk Management Standard* (CEN 2008), whose goal "is to set requirements necessary to control risks associated with the handling or storage and disposal of biological agents and toxins in laboratories and facilities" (CEN 2008:8), with a "biorisk" defined as the "combination of the probability of occurrence of harm and the severity of that harm where the source of harm is a biological agent or toxin" (CEN 2008:9).[37]

---

[35]For further information see the AG website at <http://www.australiagroup.net/>.

[36]For further information, see the CEN website at <http://www.cen.eu/cenorm/aboutus/benefits/>.

[37]"The source of harm may be an unintentional exposure, accidental release or loss, theft, misuse, diversion, unauthorized access or intentional unauthorized release."

World Health Organization

As mentioned earlier, the WHO published its most recent *Biosafety Manual* in 2004, which for the first time, contains a discussion of "biosecurity" (WHO 2004). To complement the manual, WHO published *Biorisk Management: Laboratory Biosecurity Guidance* in 2006, which attempts to "strike a balance" between longstanding biosafety practices and newer concepts of biosecurity by recommending a "biorisk management approach" to provide guidance to its member states in developing their own national approaches (WHO 2006:1). The WHO defined "biorisk" as the "probability or chance that a particular adverse event (in the context of this document: accidental infection or unauthorized access, loss, theft, misuse, diversion or intentional release), possibly leading to harm, will occur" (WHO 2006: iii).

National Regulations and Practices[38]

At present very few countries other than the United States have regulations in place governing either facilities or personnel, and for those that do—the United Kingdom (UK), France, Japan, Australia and Canada—the regulations are of recent vintage (i.e., since 2001). The exception is Germany, which has had regulations in place since 1900. Rather than attempt to describe various national practices in any detail, we offer some summary comments on the trends in various regulatory practices.

Personnel There is a wide variety among regulations addressing personnel reliability, ranging from strong local control to national registries. In the cases of local control, there is almost always a provision in the regulations that higher authorities should have access to personnel records upon request. Only the UK, Germany, and Australia appear to conduct personnel screening comparable to the SRA or security clearances. Canada has passed legislation creating the equivalent of security clearances for those working in BSL-4 laboratories, and Germany has the equivalent of security clearances for those working at the BSL-4 level already in place.

Facilities Unlike personnel, facilities are generally regulated at the national level. Such regulation takes many forms and may include requirements governing registration of specific activities. These types of regulations appear to be more common than those governing personnel, perhaps because of the prevalence of concerns with biosafety or genetically modified organisms (GMOs). In

---

[38]This section draws on the research undertaken by Dr. Robert Butera, a professor of Bioengineering and Computer Engineering at the Georgia Institute of Technology while serving as a Jefferson Science Fellow with the Department of State in 2008-2009. The material was contributed by the EO Working Group.

general, biosafety regulations are more common than those focused specifically on BSAT research.

A number of countries have tiers of regulations, requiring various levels of notification, authorization, record keeping, and so forth. Some countries require permission or licensing of facilities at a particular biosafety level, independent of the agents the facility may work with and store. Some form of registration or licensing of BSL-3 and -4 facilities is required in Germany, China, South Korea and Switzerland. There are also examples of stratification of the types of notifications and permissions required to work at various biosafety levels. In Switzerland, for example, the equivalent of BSL-2 research requires notifying the relevant authorities, while permission is required to work at the BSL-3 or -4 level. Japan has four tiers, ranging from internal record keeping at the lowest level to notification and then permission. Some activities are prohibited outright.

European countries have strict rules governing work with GMOs, which in many cases are more stringent than their rules governing pathogens. These rules tend to be focused on regulating facilities and setting standards for accounting for inventories. Since much of potential pathogen research would involve the use of recombinant DNA methods, however, the GMO regulations effectively cover a large portion of pathogen research as well.

Some countries regulate BSAT research via their regulations on biosafety risk levels. Germany, Canada, and Switzerland regulate personnel and/or facilities for all the agents designated as BSL-3 or BSL-4. Other countries regulate via lists of "select agents," which vary in length and composition, and in what special requirements they impose upon listed agents. The lists range in size from Australia's 22 to South Korea (32), France (37), Japan (51), and the UK (105).

SUMMARY

This chapter has provided background information on the origins and current operation of the Select Agent Program, additional requirements for personnel reliability and physical security that many federal agencies have applied to their own and, in some cases, their grant and contract BSAT research, and how the regulation of BSAT research is handled outside the United States. The committee drew on this information, as well as its own knowledge and experience, to develop a set of principles to guide formulation of its recommendations. Those principles are presented in the next chapter.

3

Guiding Principles for
Science and Security

This chapter proposes a set of guiding principles on how research with biological select agents and toxins (BSAT) should be viewed and conducted. These principles provide the lens through which the committee considered the specific concerns of laboratory security and personnel reliability discussed in Chapters 4 and 5.

PRINCIPLE 1: Research on biological select agents and toxins is essential to the national interest.

BSAT research is invaluable in addressing national priorities such as national security and public health. Each of the 80+ items included on the list of select agents and toxins compiled by the U.S. Department of Agriculture and Department of Health and Human Services has the potential to pose a significant threat to the health of the public or to plants or animals. Research which enhances our understanding of these agents and toxins can help diminish the threat they pose. For example, the highly successful campaign to eradicate smallpox followed from an aggressive vaccination strategy that essentially eliminated the variola virus that causes smallpox from the wild. It also enabled the United States to stockpile sufficient quantities of the smallpox vaccine to vaccinate every person in the country in the event that large-scale vaccination was ever needed. The eradication of smallpox from natural populations and stockpiling of vaccine would not have been possible without research on this dangerous select agent.

BSAT research is almost certain to have benefits both for public health and for national security. Discovery of vaccines or treatment strategies enables both preventive action and rapid response in the wake of an outbreak. Enhanced

69

technology to detect and diagnose the presence of select agents in a patient or in an environment will greatly enhance our ability to respond to—and potentially contain—the release of the agent, whether it occurs from natural infection or the deliberate use of select agents.[1] At the same time, enhanced understanding of these agents and the ability to prevent or mitigate their effects diminishes the potential impact of these agents, and thereby decreases their value as a potential terrorist weapon.

PRINCIPLE 2: Research with biological select agents and toxins introduces potential security and safety concerns.

Because BSAT materials can pose such a severe threat to health and the environment, institutions housing BSAT laboratories must do everything in their power to prevent the release of these dangerous pathogens, whether by accident or deliberate act. In fact, many of the elements of a biosafety laboratory are designed for just this purpose, but there are steps, in addition to safety procedures, that can be taken in the name of security.

The potential for use of BSAT materials in criminal or bioterrorist acts cannot be disputed. Carus (2001) has identified a number of confirmed uses of biological agents in the conduct of criminal or terrorist acts. While many involved materials not on the select agent list, several did involve the use of select agents and toxins, including *Bacillus anthracis* (anthrax), ricin toxin, *Yersinia pestis* (plague), botulinum toxin, and *Burkholderia mallei* (glanders)—which cause disease in both animals and humans. Bioagent cases are not new, extending back to the 1900s and even earlier in human history.

This emphasizes the need for robust security infrastructure designed to prevent unauthorized access to select agents facilities and to the agents themselves, as well as appropriate procedures to guard against potential insider threats, beyond those that are customary in non-select agent research. The specific security requirements should be based upon a risk analysis applicable to the particular situation and environment. Many of these procedures will also protect personnel working with select agents, as well as others, from accidental exposure in the laboratory or in the surrounding community.

PRINCIPLE 3: The Select Agent Program should focus on those biological agents and toxins that might be used as biothreat agents.

As described in more detail in Chapter 2, the listing of select agents and toxins is motivated primarily by concerns about *security*, not about *safety*.

---

[1]For example, the BioWatch program includes a network of monitoring units to detect the presence of harmful agents (see IOM/NRC 2009 for an interim report from an ongoing National Academies evaluation).

Although safety concerns are present with any human pathogen—and environmental concerns with plant and animal pathogens—inclusion of an item on the list of select agents and toxins means that it poses a security risk—namely, there is reason to believe that it could be used as a potential bioweapon.

Stated another way, the requirements for select agent research, including both personnel reliability and physical security, are motivated by—and indeed, designed to enhance—security, not safety. There is no need for security strategies unless security is the predominant consideration. Addition of unnecessary procedures will inevitably slow the development of vaccines and therapeutics—and even the public health response in the event of a biological emergency—with the unintended consequence of making the public less safe.

With these considerations in mind, the committee takes as a guiding principle that items on the list of select agents and toxins should be limited to those materials that there is reason to believe could be used as a potential biothreat agent. In addition, there should be reason to believe that enhanced security could reduce the risk of that agent being used as a bioweapon. Discussion throughout this report takes this observation as a given, and all conclusions and recommendations are based on interpreting the list of select agents and toxins—whatever its composition—as those items that pose a legitimate security risk.

PRINCIPLE 4: Policies and practices for work with biological select agents and toxins should promote both science and security.

It is common to talk about the need for a "balance" between science and security. However, this committee rejects the notion that science and security are inversely related and that there is an inherent tension between these two elements. While there may be specific circumstances when a particular action conducted to enhance security may be seen at odds with the conduct of scientific research, the two aims are not fundamentally opposed. Rather, the policies that support science and those that promote security operate in different but overlapping spheres. The goal for this report and for the community is to optimize the mix of policies that promotes both high-quality science and appropriate security.

PRINCIPLE 5: Not all laboratories and not all agents are the same.

As described in Chapter 2, there is significant diversity among those institutions that conduct BSAT research. Academic, government, and private-sector select agent laboratories span a wide range in mission, size, type of research, level of activity, and many other characteristics. Colloquially, "if you've seen one lab, you've seen one lab." While this phrase may downplay the common elements found in multiple environments, it is critical to acknowledge that each setting has its own unique set of circumstances and issues. This speaks to the continued need for site-specific risk assessments and security approaches with enough flexibility to both promote security and safety and properly address the characteristics specific to that institution.

It is also critical to recognize that not all select agents and toxins are the same, nor do they all pose the same risks. This speaks to the need for a graduated set of risk-based policies and practices that adequately addresses the specific needs of each facility and program, as opposed to a one-size-fits-all approach that is not optimized for any given setting.

PRINCIPLE 6: Misuse of biological materials is taboo in every scientific community.

Despite diversity in the facilities and agents used in BSAT research, there is a shared ethos throughout the scientific community that *any* misuse of biological materials is taboo. The intentional use of disease to cause harm is contrary to the fundamental goals of the life sciences to contribute to the welfare of all living things and to the safety of the environment. The use of biological materials as a weapon is simply not accepted as legitimate by almost every scientist and country: no reputable scientist or scientific organization would purposely perpetrate an action that puts the public at risk.[2]

These principles have been enshrined in numerous international agreements (see discussion in Chapter 2), and they are seen to be absolute. The scientific community has a responsibility for helping to make sure that the misuse of biological materials remains taboo. Individual scientists cannot be expected to do the impossible, so scientists cannot be expected to *ensure* that the knowledge they generate will never contribute to the advancement of biowarfare or bioterrorism. But scientists can and should be expected to take reasonable steps to reduce the risk that their science, and the results of life sciences research more generally, could be misused.

Scientists, and the scientific community more broadly, can exercise this responsibility in many ways. Individual awareness is important, as is education and training to create and maintain a culture of trust and responsibility that is central to sustaining good scientific conduct. Professional societies play an essential role, through their own training programs, by promoting responsible behavior through codes of conduct, and by supporting policies and practices that can help reduce the risks of misuse.

---

[2]To be sure, there may be some discussion about which experiments are risky, but there is general agreement that the science itself should not be misused.

4

Issues Related to Personnel Reliability

INTRODUCTION

For those concerned about the security of laboratories conducting research with biological select agents and toxins (BSAT), personnel issues are among the most difficult and controversial. Many of the proposals and new policies that followed from the July 2008 conclusion by the Federal Bureau of Investigation (FBI) that Bruce Ivins, a longtime employee at the U.S. Army Medical Research Institute of Infectious Disease, was responsible for the 2001 anthrax attacks focused on how to be more proactive in order to prevent another such incident by identifying individuals who may pose a threat before they can act. For example, much of the time and attention of the Executive Order (EO) Working Group on Strengthening the Biosecurity of the United States and its public consultations and site visits were devoted to the challenge of how the nation could guard against such threats. In fact, several of the reports offered in response or related to the EO process focused only on personnel issues (NSABB 2009; AAAS 2009; Leduc et al. 2009). This committee's charge includes both personnel reliability issues and physical security. But because current practices and the prospect of additional measures related to personnel assurance have caused so much anxiety about the impact of the measures on the ability to attract and retain high-quality research and technical personnel and conduct the best science, the committee has devoted an entire chapter to these specific issues.

The first part of this chapter discusses *screening*, that is, the process of identifying whether or not someone should be eligible to have access to BSAT materials. The second part of the chapter, recognizing that individuals accused or convicted in a number of major U.S. terrorism and espionage cases had already passed the screening phase, addresses how one might *monitor* employee behavior and performance and *manage* the workplace to reduce the risk of an insider either carrying out thefts or sabotage or acting to assist others.

SCREENING

Introduction

Personnel screening seeks to identify individuals who may pose a potential security risk as early as possible, ideally prior to hiring. Identifying security risks can be considered part of the broader challenge of hiring competent, trustworthy, and reliable employees, and most organizations have a selection procedure to identify education and training, competencies, aptitude, and experience among potential employees.[1] Many private- and public-sector organizations also conduct background checks to identify (in)appropriate actions or to assess personal qualities that are considered desirable or necessary for effective job performance. As discussed in this section, screening for security risks poses special issues.

The current screening process for individuals to work in facilities conducting BSAT research is based on identifying any of a set of disqualifying behaviors/activities that would automatically and permanently deny a person access (see Chapter 2 for additional details). Most of the policy discussions about the current Security Risk Assessment (SRA) screening process focus on four issues:

1. the adequacy of the information used to assess individual applicants;
2. the necessity to make changes in the types of information collected as part of the background checks;
3. the need to make changes in the way the current SRA process makes decisions about granting access; and

---

[1] "Selection procedures refer to any procedure used singly or in combination to make a personnel decision including, but not limited to, paper-and-pencil tests, computer-administered tests, performance tests, work samples, inventories (e.g., personality, interest), projective techniques, polygraph examinations, individual assessments, assessment center evaluations, biographical data forms or scored application blanks, interviews, educational requirements, experience requirements, reference checks, background investigations, physical requirements (e.g., height or weight), physical ability tests, appraisals of job performance, computer-based test interpretations, and estimates of advancement potential. These selection procedures include methods of measurement that can be used to assess a variety of individual characteristics that underlie personnel decision making" (SIOP 2003:3). "The essential principle in the evaluation of any selection procedure is that evidence be accumulated to support an inference of job relatedness. Selection procedures are demonstrated to be job related when evidence supports the accuracy of inferences made from scores on, or evaluations derived from, those procedures with regard to some important aspect of work behavior (e.g., quality or quantity of job performance, performance in training, advancement, tenure, termination, or other organizationally pertinent behavior)" (SIOP 2003:4).

4. the possibility of adding additional forms of screening, in particular various types of psychological tests.

It is vital to acknowledge the formidable challenges posed by screening individuals for potential security concerns. The proportion of the population of job candidates who represent true security risks is unknown, but likely to be very small. This low base rate makes it difficult to detect true threats because "screening in populations with very low rates of the target transgressions (e.g., less than 1 in 1,000) requires diagnostics of extremely high accuracy" (NRC 2003:5), and these do not exist for the problems we are trying to address (or for many others). There is no way to escape the risk that good candidates will be screened out in order to detect a small number of people who pose genuine threats to security. This is not a new issue and, as discussed in Chapter 2, the U.S. government attempts to address this dilemma through a number of approaches aimed at assuring personnel reliability.

Efforts at screening for rare individuals or behaviors will therefore inevitably struggle with concerns about either failing to identify someone who has the disqualifying background or behavior or identifying someone as having disqualifying background or behavior when she or he does not. These two concerns are inversely related: the more one tries to avoid letting a security risk get through the screening, the more one increases the number of innocent individuals who will "fail" the test. The 2003 National Research Council (NRC) study *The Polygraph and Lie Detection* illustrates the difficult trade-offs facing policymakers with the example of a polygraph screening exam with an accuracy index of 0.90 for a hypothetical population of 10,000 government employees that includes 10 spies:

> If the test were set sensitively enough to detect about 80 percent or more of deceivers, about 1,606 employees or more would be expected [to] "fail" the test; further investigation would be needed to separate the 8 spies from the 1,598 loyal employees caught in the screen. If the test were set to reduce the numbers of false alarms (loyal employees who "fail" the test) to about 40 of 9,990, it would correctly classify over 99.5 percent of the examinees, but among the errors would be 8 of the 10 hypothetical spies, who could be expected to "pass" the test and so would be free to cause damage. (NRC 2003:6)

In addition to the general dilemma of such trade-offs, the impact of unnecessarily excluding someone who does not introduce a security risk poses a special problem for the technical and research personnel in the BSAT workforce. If there is a large pool of potentially qualified applicants, a manager could decide that she or he can "afford" to incorrectly exclude someone who is in fact qualified because there are many others from whom to choose. (Even if the employer is not affected, "failing" the test could have harmful consequences for the innocent individual involved, especially if there is a risk of any lasting career impact.) But tangible costs may be incurred when highly skilled workers are incorrectly excluded from consideration. Because there may be a relatively small number of qualified candidates, especially for senior research positions, turning away a good candidate will entail at least the costs of finding a replacement, if one even exists. Moreover, SRA screening will only take place *after* an individual has been selected for other reasons. Even graduate students considering work with BSAT materials have already been selected for advanced study because of other, desirable characteristics and have undergone significant periods of training.

In addition, difficulties during the screening process may also create a disgruntled applicant who may continue to be part of a relatively small specialized research community. Experts in personnel screening have long been concerned with the challenge that a system applicants find too intrusive or unfair could make even successful applicants feel the selection process is unjust, creating negative feelings or attitudes that could ironically contribute to someone's becoming disgruntled and potentially susceptible to the very behavior screening is intended to prevent (Murphy 2009). Although there does not appear to be clear empirical evidence that screening systems actually affect the subsequent behavior of selected applicants (Sackett and Lievens 2008:438), the perception of the research community should be considered in designing screening procedures for those working with BSAT materials.

Finally, the Society for Industrial and Organizational Psychology (SIOP) has recognized a potential negative consequence from the use of testing—creating complacency or a false sense of security—that could apply to any form of screening. Testing may prompt institutions to relax other procedures, for example to reduce theft, because they believe the threat to have been eliminated:

> An organization that introduces an integrity test to screen applicants may assume that this selection procedure provides an adequate safeguard against employee theft and will discontinue use of other theft-deterrent methods (e.g., video surveillance). In such an instance, employee theft might actually increase after the integrity test is introduced and other organizational procedures are eliminated. Thus, the decisions subsequent to the introduction of the test may have had an unanticipated, negative consequence on the organization. (SIOP 2003:7)

With this brief introduction to the challenges of screening for security risks, the next two sections consider (1) the current SRA and whether changes to either the disqualifying background/activities or the operation of the process is warranted; and (2) whether other screening methods, in particular testing, would add to the confidence that one could identify problematic potential employees.

Identifying Individuals with Backgrounds or Activities That Could Pose a Risk

Issues with the Current SRA

The committee considered the appropriateness of current criteria included in the SRA as disqualifying factors[2] and whether changes should be made to the implementation of the screening process (see Chapter 2 for a description of the current SRA). The very small number of rejections and appeals reported by the Select Agent Program—192 rejections out of a total of 31,349 applications processed and 58 appeals, of which 22 resulted in the denial being overturned[3]—can be interpreted either that the screening is not restrictive enough, allowing potential risks to gain access to BSAT facilities, or as effective institutional pre employment screening that weeds out those ineligible for access to BSAT materials prior to the SRA process. Without baseline information about the actual number of high-risk candidates, there is no empirical basis for using these rejection rate data to infer that the process is flawed.

Before offering its assessment of the current SRA, the committee notes the need for the Select Agent Program to clarify what constitutes some of the background/activities considered disqualifying factors. The public consultations held by both the National Science Advisory Board for Biosecurity (NSABB) and the EO Working Group revealed a substantial lack of understanding of how issues related to sexual orientation and mental health are addressed by the SRA. This confusion appears to be increasing the concern of the research community about whether the criteria are appropriate. Contrary to the expressed concern, however, individuals who are separated from the armed services as a result of the current "don't ask, don't tell" policy or because of personality disorders that, with proper medication and/or treatment, could permit effective

---

[2]Under the USA PATRIOT Act a "restricted person," that is, someone permanently disqualified to work in a BSAT facility, is under indictment for a crime punishable by imprisonment for a term exceeding one year, has been convicted in any court of a crime punishable by imprisonment for a term exceeding one year, is a fugitive from justice, is an unlawful user of any controlled substance (as defined in section 102 of the Controlled Substances Act (21 USC 802)), is an alien illegally or unlawfully in the United States, has been adjudicated as a mental defective or has been committed to any mental institution, is an alien (other than an alien lawfully admitted for permanent residence) who is a national of a country that has repeatedly provided support for acts of international terrorism, or has been discharged from the Armed Services of the United States under dishonorable conditions. The Bioterrorism Preparedness Act added prohibitions for any person reasonably suspected by any federal law enforcement or intelligence agency of committing a crime specified in 18 USC 2332b(g)(5) [a "federal crime of terrorism"], having a knowing involvement with an organization that engages in domestic or international terrorism (as defined in 18 USC 2331) or with any other organization that engages in intentional crimes of violence, or being an agent of a foreign power (as defined in 50 USC 1801).

[3]These data were provided by Julia Kiehlbauch from the Agriculture Select Agent Program on behalf of the Animal and Plant Health Inspection Service, Centers for Disease Control and Prevention, and FBI.

functioning in a nonmilitary setting, would *not* receive dishonorable discharges unless they committed offenses that resulted in conviction by a court marshal.[4] The restriction on an individual who "has been adjudicated as a mental defective or has been committed to any mental institution" also raised concern in the public consultations. However, the SRA does not affect people who, for example, are: (1) suffering from problems such as bipolar disorder or forms of depression; (2) voluntarily undergoing mental health treatment; or (3) have been voluntarily hospitalized for mental health problems in the past.[5] The committee believes the Select Agent Program can help reduce these concerns by providing more specific guidance about what is meant by these terms and perhaps by including clarification on the SRA form itself.[6]

In making its assessments, the committee considered how the SRA compares with other basic security and suitability screening carried out by the federal government. The broader context is important for understanding the committee's conclusions and recommendation; it seems reasonable, for example, to ask how the SRA compares to a process that enables 2.4 million people to have access to various levels of classified information (GAO 2009b).[7] Although the committee did not have time to conduct a thorough review of processes in all parts of the government that affect scientists, there could be important lessons or cautionary tales from the long experience in several areas, such as the nuclear weapons laboratories or the National Security Agency, which carries out research in cryptography. As described in Chapter 2, the committee did consider personnel security requirements beyond the SRA that various federal agencies have adopted for their staff and, sometimes, for their contractors and grantees.

Potential Changes to the Current SRA

Of the various changes to the current SRA discussed in the public consultations and assessments of the program to which the committee had access (e.g., NSABB 2009; DSB 2009), four particular issues garnered enough attention that the committee decided to address them.

Adding Additional Databases to the Current Screening   One question that has arisen in policy discussions is whether the FBI, which carries out the back-

---

[4]See a report by the Congressional Research Service (Burrelli 2009) for discussion of discharge policies related to "don't ask, don't tell."

[5]However restrictive the definition, it still goes against much current thinking among medical health professionals not to treat someone who has suffered from a severe mental disorder and been successfully treated as permanently disabled or, in this case, permanently ineligible.

[6]The NSABB also recommends clarifying the mental health issues in its report (NSABB 2009:12).

[7]This figure, which does not include individuals who work in some areas of national intelligence, has declined from an estimated 3.2 million people who held security clearances in 1993.

ground checks for the SRA, is taking advantage of all the appropriate databases to which it has access. (Chapter 2 contains a list and discussion of the databases currently being used.) Discussion among the members of the committee, including several who have experience with the databases used for this type of screening, consultation with a number of outside experts, including in a public session during the committee's second meeting devoted to federal security and suitability screening practices, and public discussion about these issues suggested that the SRA may be consulting even more databases than those in the routine federal security clearance process. Although there may be specialized databases held by other agencies to which the FBI would not have access, information available to the committee suggests that the databases used for the current SRA are equivalent or comparable to those used for most other federal screening processes. **The committee concluded that the databases being used in the SRA are consistent with current U.S. government practices in determining the eligibility of persons to have access to classified and proprietary information and sensitive sites and are adequate for assessing whether applicants possess disqualifying background/activities.**

Adding a Mandatory Drug Test The current SRA addresses *past* use of illegal drugs only through database checks that identify anyone convicted of a crime carrying a potential prison term greater than one year, which would include drug-related crimes. By contrast, the general application form for a federal security clearance (Form SF86) maintained by the Office of Personnel Management (OPM) asks about illegal use of any controlled substances or prescription drugs since the age of 16 or within the past seven years, whichever is shorter. A number of federal agencies—and private firms as well—have concluded that experimentation with illegal drugs is so common among U.S. young adults that the agencies do not consider that admission of past use necessarily makes someone ineligible. Acknowledgment of past illegal use of drugs is not automatically disqualifying in these cases, although any applicant who did admit to past use could expect to be questioned further. In any case, agencies would terminate an employee who continued that use once on the job.

As opposed to past and noncontinuing use, the current SRA policies with regard to current use of illegal drugs are consistent with the broader federal approach. Public Law 110–181, Section 3002, prohibits any officer or employee of a federal agency, including active-duty military and federal contract employees, from being granted, or maintaining continued eligibility for, a security clearance if they are an unlawful user of a controlled substance or an addict. No waivers are permitted, which is consistent with the SRA.[8] The SRA assesses current use through a question on the application form (see Appendix D), but the issue is

---

[8]Information provided by electronic communication with OPM and the Office of the Director of National Intelligence, August 6, 2009.

whether to add a mandatory test to verify an applicant's statement that he or she is not using illegal drugs.

In addition to potential security risks, use of illegal drugs could be regarded as a safety issue. Successfully passing a drug test could also be considered a sign of reliability or evidence of respect for the law. This type of testing is becoming more common in industry and government, but not in academia. Routine drug testing could also be part of ongoing monitoring of employees.

The committee concluded that there was insufficient information to say that routine or random drug testing would significantly reduce the risk of an insider threat. The committee noted, however, that use of illegal drugs provides insight into a person's judgment and reliability, which are critical attributes for those with access to highly pathogenic infectious agents. If the select agent list is stratified, consideration could be given to adding a mandatory drug test for those who would have access to agents in the highest risk group.

Adding a Credit Check or Financial History   An obvious omission from the current SRA is querying an applicant's financial and credit history. At least some consideration of credit history is common in many sectors as part of pre-employment screening and standard practice in federal security clearance and suitability investigations. In most cases, however, the issue is not one of an individual's level of debt per se, but whether spending patterns provide a means to assess judgment and reliability and possible vulnerabilities. This information would be a logical element for inclusion in ongoing assessment and monitoring of employees, which is discussed later in the chapter.

A major reason for considering addition of financial information is that greed or susceptibility to bribery has been found to be a factor, in some cases, in the decision to become an accomplice to those undertaking illegal acts. Most espionage cases during the end of the Cold War, for example, involved spies acting out of economic rather than ideological motivation (Herbig and Wiskoff 2002). However, a 2008 analysis of data collected by the Defense Personnel Security Research Center on about 170 U.S. citizens who committed espionage between 1947 and 2007 showed a more complicated picture:

- "Since 1990, money has not been the primary motivation for espionage. While getting money was the sole motive for 47 percent of the first cohort[9] and 74 percent for the second cohort, since 1990 only 7 percent (which represents one individual) spied solely for the money. Money remained one of multiple motives in many recent cases as well" (Herbig 2008:ix).
- Since 1990, 35 percent of the spies that were apprehended were naturalized citizens (as compared to 80 percent native-born before that time), 58 percent had "foreign attachments" (relatives or close friends overseas),

---

[9]The analysis divides the cases into three cohorts: those that began their spying between 1947 and 1979; those that began between 1980 and 1989; and those that began in or after 1990.

and 50 percent had foreign business or professional connections, with the result that, whereas "divided loyalties" were the sole motive for less than 20 percent prior to 1990, since 1990 that number has increased to 57 percent (Herbig 2008:vi-vii).

- "Since 1990, the proportion of American spies demonstrating allegiance to a foreign country or cause more than doubled to 46 percent compared to the 21 percent in the two earlier cohorts, reinforcing the sense that globalization has had a noticeable impact and that the influence of foreign ties has become more important since 1990" (Herbig 2008:x).

A formidable barrier to adding financial history as a consideration to the current SRA, however, is that there is no clear indicator or threshold from which to base a decision about whether someone should be automatically disqualified. Any assessment would need to be appropriate for the particular segment of the applicant pool. For example, many students and those in training will have student-loan debts, some of them very heavy. Trainees are also likely to have relatively low salaries, especially relevant to their educational attainment. Scientists from outside the United States may not have a credit history that would permit them to obtain credit cards and other normal measures of financial responsibility. Judgment would inevitably be required, and the current SRA process does not include that kind of discretion.

The committee concluded that the difficulties of establishing a meaningful baseline make adding credit or financial history to the current SRA screening process too challenging. In any event, signs of sudden, unexplained affluence or evidence of irresponsible financial behavior would be appropriate to consider as part of the process of monitoring employees' behavior.

Adding an Adjudication Process The current practice of automatic and permanent denial of eligibility for anyone who reveals or is found to have any disqualifying factor has raised concern. The current SRA system has no statute of limitations on disqualification: it does not matter how long ago the offense was committed. There is also no consideration of extenuating circumstances. The only appeal is to permit correction of factual errors.

By contrast, information collected under other current federal suitability and security screening is subject to an adjudication process, whereby issues such as how long ago the offense occurred, whether recent behavior shows positive or negative trends, and mitigating circumstances are taken into account to determine whether to grant access to protected information. Guidelines for making these determinations are available and periodically reviewed and updated (White House 2005). The appeal process also can take these factors into account in assessing whether a decision to deny access was justified.

The committee considered whether the SRA process should more closely mirror the security screening process by introducing adjudication to provide an opportunity for considering the circumstances of a disqualifying offense. Although the reform measures undertaken in response to EO 13467 are reducing the processing time for security and suitability investigations, introducing judgment into the current process would almost certainly make the screening longer and more expensive (see Chapter 2 for discussion about security clearance investigations). The research community has already expressed concern about the length of time needed to clear personnel for access, so adding to the time would likely be perceived as a further inconvenience. Moreover, the small number of exclusions suggests that adjudication need not be incorporated in all cases, but only as part of the appeal process.

The committee concluded that the questions raised about the current automatic and permanent disqualifications were sufficiently serious that it would be worthwhile to change the system to incorporate a broader appeal process more aligned with personnel security practices already in place across the government.

Recommendation

The committee's conclusions with regard to potential changes to the SRA are conditional because we believe the appropriateness of additional measures, in some cases, depends on whether or not the recommendation in Chapter 5 to stratify the list of select agents and toxins by risk groups is adopted. A stratified list, which presumably would restrict the highest level of security measures to a smaller set of agents and toxins, would also dictate a stratified SRA that could add additional requirements only to those who would work with those agents and toxins in the most stringent risk group.

> **RECOMMENDATION 5: The current Security Risk Assessment screening process should be maintained, but the appeal process should be expanded beyond the simple check for factual errors to include an opportunity to consider the circumstances surrounding otherwise disqualifying factors.**

Identifying Potential Insider Threats through Testing

Introduction

Policy discussions have included the issue of whether to require more extensive testing and evaluation of applicants to work with BSAT materials, perhaps as part of a formal Personnel Reliability Program. Some government agencies and private entities, including academic institutions, have considered undertaking additional screening using psychological or psychophysiological tests. This section discusses the types of testing available and what is known about their appropriateness and effectiveness for these purposes.

Given the various definitions of what constitutes an insider threat (see Chapter 1), at least two different types of problems need to be addressed when individuals are screened to identify those potentially posing a threat. One set of problems arises in determining the normal range of adult personality; persons outside of this range would be identified as those who either might attempt deliberate deception or those who might be susceptible to corruption or recruitment to aid in the theft of materials or acts of sabotage. Another set of problems involves identifying individuals suffering from a range of serious personality disorders that might lead to their using BSAT materials to deliberately cause harm or assist others in doing so. In making these broad and admittedly inexact distinctions, we are not addressing individuals who might provide unwitting aid through a lack of awareness and those who might be subject to coercion that background checks can help identify.

There is an extensive literature on approaches to identifying insider threats, including from terrorists (see discussion below). There is also extensive experience from government and the private sector using various types of tests for screening purposes. Currently available tests fall into two broad categories. Polygraph exams are the best known and most commonly used example of *psychophysiological tests,* which rely on assessing the body's physiological responses. *Psychological tests* include both "normal range" testing and tests that measure possible aberrant or psychopathological traits.

Polygraph Testing

Polygraph testing is described here because it is used by some government agencies for national security screening—including some who may conduct BSAT research. A polygraph is an instrument that measures and records several physiological responses such as blood pressure, pulse, respiration, breathing rhythms, body temperature, and skin conductivity while the subject is asked and answers a series of questions; it is based on the theory that false answers will produce distinctive measurements that a skilled examiner will be able to recognize and interpret. The polygraph is used in a variety of settings for (1) investigation of specific incidents—such as in law enforcement situations, (2) evaluation of current employees, and (3) assessment of prospective employees.

The NRC produced a report on *The Polygraph and Lie Detection* in 2003 at the request of the Department of Energy, which had begun using polygraph testing for some personnel at nuclear weapons laboratories in response to the alleged spy activities of Wen Ho Lee at Los Alamos National Laboratory. In addition to its extensive review of polygraph testing, the study also examined alternatives to the polygraph that might provide other means to detect deception in job applicants, current employees, or in investigations (NRC 2003). This committee found the 2003 report useful for its work.

The study noted an important distinction between the use of polygraph testing in the context of a specific investigation (e.g., whether a person was or was not involved in a particular incident of wrongdoing) versus broad use to assess risk for future involvement in wrongdoing. The study found that polygraph testing in such specific investigations could produce accurate results at "rates well above chance, though well below perfection" (NRC 2003:4), for those not trained in deceptive tactics. Polygraphs used for investigations of a particular occurrence are quite focused, concentrating on one event, and retrospective, so that precise true/false questions of fact are the focus of the exam. The study found that polygraphs were far less reliable for other purposes; for example, in national security screening, the exam covers a range of past behaviors, which might include ambiguous or speculative situations where the examiner and the subject do not have the same picture of a situation, even when asking true/false questions. The subject's responses are then the basis for making inference to his or her future behavior. The polygraph study committee concluded that:

> Available evidence indicates that polygraph testing as currently used has extremely serious limitations in such screening applications, if the intent is both to identify security risks and protect valued employees. Given its level of accuracy, achieving a high probability of identifying individuals who pose major security risks in a population with a very low proportion of such individuals would require setting the test to be so sensitive that hundreds, or even thousands, of innocent individuals would be implicated for every major security violator correctly identified. The only way to be certain to limit the frequency of "false positives"[10] is to administer the test in a manner that would almost certainly severely limit the proportion of serious transgressors identified. (NRC 2003:6)

A more recent NRC study on the use of newer technologies to detect deliberate falsehoods found that, "to date, insufficient, high-quality research has been conducted to provide empirical support for the use of any single neurophysiological technology, including functional neuroimaging, to detect deception" (NRC 2008b:4).

The 2003 polygraph study committee recognized that polygraphs might have other uses, even if they are not accurate—such as deterring poor security risks from applying in the first place or making employees more likely to confess violations that they believed would be detected by polygraphs. These effects

---

[10]In the context of this type of testing, a "false positive" is the opposite of the typical meaning in areas such as medical diagnostics. Here, a false positive refers to someone being identified as truthful when he or she is in fact deceptive.

could be obtained whether or not the polygraph was accurate in detecting a falsehood and might, in fact, account for why some federal agencies continue to use polygraphs.

Normal Range Testing: Integrity Tests

Normal range psychological testing covers a wide variety of assessment strategies. "Integrity tests" include a variety of instruments used to assess attitudes and experiences related to an individual's honesty, dependability, trustworthiness, reliability, and pro-social behavior. These are the tests most commonly used to identify potentially counterproductive workplace behavior. According to SIOP, integrity tests "typically ask direct questions about previous experiences related to ethics and integrity OR ask questions about preferences and interests from which inferences are drawn about future behavior in these areas. Integrity tests are used to identify individuals who are likely to engage in inappropriate, dishonest, and antisocial behavior at work" (SIOP 2009a). A survey conducted in 2001 by the American Management Association (AMA), which reflects practices at the large organizations that are AMA members rather than the population of all U.S. employers, found that 29 percent of employers surveyed use one or more forms of psychological measurement or assessment, which would also include personality tests (SIOP 2009b).

Although integrity testing was originally developed to detect dishonesty without having to make use of polygraph tests—with a particular focus on reducing theft—its applications have expanded over the years to cover broader concepts of theft (e.g., "time theft" through absenteeism, low productivity) and other types of counterproductive workplace behavior (Berry et al. 2007).[11] Reviews of published research concerning integrity testing suggest such testing can produce valid predictions of potential counterproductive behavior (Sackett and Harris 1984; Sackett et al. 1989; Sackett and Wanek 1996; NRC 2003; Berry et al. 2007). Integrity tests have also been shown to predict job performance, which is not surprising: "employees who engage in a wide variety of counterproductive behaviors are unlikely to be good performers" (NRC 2003:173).

---

[11]Integrity tests fall into two general categories. "'Overt' integrity tests commonly consist of two sections. The first is a measure of theft attitudes and includes questions pertaining to beliefs about the frequency and extent of theft, punitiveness toward theft, ruminations about theft, perceived ease of theft, endorsement of common rationalizations for theft, and assessments of one's own honesty. The second involves requests for admissions of theft and other wrongdoing. Commonly used tests of this type include the Personnel Selection Inventory (PSI), the Reid Report, and the Stanton Survey. 'Personality-oriented' measures are closely linked to normal-range personality devices, such as the California Psychological Inventory. They are generally considerably broader in focus than overt tests and are not explicitly aimed at theft. They include items dealing with dependability, Conscientiousness, social conformity, thrill seeking, trouble with authority, and hostility. Commonly used tests of this sort are the Personnel Reaction Blank, the PDI Employment Inventory (PDI-EI), and Reliability Scale of the Hogan Personality Series" (Berry et al. 2007:271-272).

Most relevant to this report is how useful integrity testing is for detecting potential insider threats. As the NRC study of polygraphs concluded, "There is no literature correlating the results of these tests with indicators of the more specific kinds of counterproductive behavior of interest in national security settings" (NRC 2003:173). Because counterproductive behaviors studied are often correlated (i.e., a person willing to engage in one is more likely to also engage in another), one might posit that there would be a relationship to other specific counterproductive behaviors that have not yet been studied. It is not clear, however, whether this applies in the context of bioterrorism: how likely would someone who could be recruited to steal equipment or other materials from a lab as an accomplice in a "normal" theft be to steal BSAT materials when it would presumably be apparent this was being done for purposes of terrorism or sabotage?

Personality Assessment Tools

Concerns about insider threats also include those who are suffering from mental disorders severe enough to potentially cause them to commit illegal acts. In this section, we address the issue of whether such problems could be identified at the point of hiring; the challenge of identifying and responding to such problems once someone is already working in a facility is addressed later in the chapter.

A number of standardized tests have been developed to aid in the effort to identify employees who suffer from psychopathology or personality disorders. The original personality assessment tests were developed during World War I by the U.S. military for screening draftees, and such standardized tests are commonly used in a number of government and private settings (Butcher et al. 2006). A number of high risk or sensitive occupations, such as the military and the police, make use of such tests; for example, the use of a clinical test instrument is required by law for candidates for jobs as law enforcement officers in 50 percent of the states (Cullen et al. 2003).

One of the most widely used clinical personality assessments is the Minnesota Multiphasic Personality Inventory (MMPI). It is used in nonclinical settings to identify a range of psychopathologies and to assess persons who are candidates for high-risk public safety positions, such as nuclear power plant personnel, police officers, firefighters, pilots, and air-traffic controllers. Originally developed in the 1930s, it has been refined over the years and has been the subject of extensive research.[12] Results are interpreted by examining the relative

---

[12]See, for example, Cullen et al. (2003) and Butcher et al. (2006).

elevation of factors compared to the various reference groups studied.[13] Other frequently used assessment tools are the Millon Clinical Multiaxial Inventory-III (MCMI-III) and the Personality Assessment Inventory (PAI). It is considered good clinical practice not to rely on one test exclusively, and judgments about any individual are more reliable when tests are used in combination and test results are supported by other methods of assessment.

A key question is how well the many standardized tests developed to assess personality are able to identify potential problem employees. And even if the tests are effective for this purpose, one then needs to ask whether the traits they identify are related to the specific problem one is trying to solve: excluding potential insider threats and terrorists from the laboratory. Unfortunately, whatever clinical diagnostic instrument one might choose to screen for potential insiders and possibly a terrorist, the test will be vulnerable to the same difficulties that beset polygraphs and integrity testing when trying to identify rare behaviors.

There is little evidence that potential bioterrorists are more likely to come from among the ranks of those with a given specific psychopathology than those motivated by some other reason, such as commitment to a cause that uses terrorism or those who would undertake terror for financial gain. In fact, research suggests that, however abhorrent their actions may be to most people, "the outstanding common characteristic of terrorists is their normality" (Crenshaw 1981:390). An extensive recent review of the research on the "psychology of terrorism" for one of the U.S. intelligence agencies concludes that:

> Research on the psychology of terrorism has been nearly unanimous in its conclusion that mental illness and abnormality are typically not critical factors in terrorist behavior. Studies have found that the prevalence of mental illness among samples of incarcerated terrorists is as low or lower than in the general population. Moreover, although terrorists often commit heinous acts, they would rarely be considered classic "psychopaths." Terrorists typically have some connection to principles or ideology as well as to other people (including other terrorists) who share them. Psychopaths, however, do not form such connections, nor would they be likely to sacrifice themselves (including dying) for a cause. (Borum 2004:34-35)

---

[13]There is also a large body of literature on the use of the MMPI with various native language groups; this question came up in the NSABB public consultations because of the number of foreign nationals or green card holders and naturalized citizens potentially involved in BSAT research. The ongoing academic debate over the validity of newer versus older versions of the MMPI is beyond the scope of this report (see, for example, the special issue of the *Journal of Assessment* in 2006 (Vol. 87, Issue 2) devoted to the topic).

Summing Up

This brief review demonstrates the variety of tests that might be considered as part of a screening program to identify those individuals who pose a potential insider threat before they enter the laboratory. **The committee concluded that there is no "silver bullet," that is, no single assessment tool that can offer the prospects of effectively screening out every potential terrorist. Although it can be appropriate for organizations to employ integrity testing and clinical personality assessments as part of screening to serve other purposes, the committee reached the same conclusion concerning polygraph testing as was reached by another NRC committee that applies even more broadly, namely to its use in security screening: "Polygraph testing yields an unacceptable choice for...employee security screening between too many loyal employees falsely judged deceptive and too many major security threats left undetected. Its accuracy in distinguishing actual or potential security violators from innocent test takers is insufficient to justify reliance on its use in employee security screening in federal agencies" (NRC 2003:6).**

MONITORING AND MANAGEMENT TO ACHIEVE A SAFE AND SECURE RESEARCH ENVIRONMENT

Introduction

The current SRA process is built upon screening an array of databases for certain disqualifying behavior/background factors. Once an individual is cleared, certification is in effect for five years. However, the FBI continues to monitor cleared individuals using selected databases; the FBI also receives automatic notices in some instances, for example, when an individual is arrested and fingerprinted (NSABB 2009:3).

Sustained database monitoring can help identify that a cleared individual has incurred at least some of the disqualifying factors that would make him or her ineligible to work with BSAT materials. But the process cannot be expected to address all disqualifying factors or, perhaps more importantly, all significant issues and personal changes that could occur in an individual's life during the five-year period of certification, including those that could potentially result in his or her becoming a security risk. The conclusion that one should not rely exclusively on screening to identify potential insider threats before hiring makes this recognition both important and troubling. It implies that policymakers will not have easy or easily measurable remedies for the concerns about personnel reliability. More importantly—and positively—it suggests that efforts to ensure personnel reliability will have to come from the laboratories where BSAT research is being conducted, in the form of increased engagement by managers and staff. To appreciate the potential of such engagement, it is necessary to address a persistent belief that affects how the impact of monitoring laboratory personnel is commonly viewed.

Dispelling a Myth about Spontaneous Action

Over the years, an extensive literature has accumulated on preventing insider threats, covering a wide range of types, from espionage, fraud, corruption, and misuse of information technology or other systems containing secure or proprietary information, to threats and acts of violence that include the workplace and schools (Turner and Gelles 2003; Herbig 2008; Fein and Vossekuil 2009; Brant and Gelles 2009). The research includes many case studies of terrorism and some bioterrorism incidents in particular.[14]

One important lesson from this research is that, even in circumstances where one might assume an individual would attempt to conceal his or her malevolent intent in order to escape detection, *in many cases there will be signs or signals that something is wrong prior to an event*. Those cases in which an individual's action is genuinely spontaneous are rare. Most people follow a psychological path from idea to action and give signals along the way (Fein et al. 1995; Fein and Vossekuil 2009; see Borum 2004 for a discussion focused on terrorists). The warning signs occur often enough that it is reasonable to believe that active, sustained monitoring and management could detect many of them and provide the basis for prevention (Turner and Gelles 2003; Fein and Vossekuil 2009). No system can guarantee success in preventing an illegal act, but the results of the research on insider threats just discussed are hopeful. The research also suggests that training people to watch for and recognize the warning signs is essential and that, in the absence of such training, these signs are likely to be missed (Cascio 2009). This leads directly to one of the committee's most important recommendations.

RECOMMENDATION 1: Laboratory leadership and the Select Agent Program should encourage and support the implementation of programs and practices aimed at fostering a culture of trust and responsibility within BSAT entities. These programs and practices should be designed to minimize potential security and safety risks by identifying and responding to potential personnel issues. These programs should have a number of common elements, tailored to reflect the diversity of facilities conducting BSAT research:

---

[14]See Borum (2004) for a wide-ranging review of the state of the art of research on the psychology of terrorism, including issues relevant to the question of warning signs. See Carus (2001) and Tucker (2000) for case studies of bioterrorism. It is important to note that the various projects involving case studies of bioterrorism do not necessarily reach the same conclusions, in particular about the risks of terrorists turning to bioterrorism.

- **Consideration should be given to including discussion of personnel monitoring during (1) the initial training required for all personnel prior to gaining access to BSAT materials and annual refresher updates and (2) safety inspections to obtain a more complete assessment of the laboratory's ability to provide a safe and secure research environment.**
- **More broadly, personnel with access to select agents and toxins should receive training in scientific ethics and dual-use research. Training should be designed to foster community responsibility and raise awareness of all personnel of available institutional support and medical resources.**
- **Federal agencies overseeing and sponsoring BSAT research and professional societies should provide educational and training resources to accomplish these goals.**

The remainder of the chapter describes how the most important parts of the recommendations can be implemented, including:

- The types of education and training needed to foster a culture of responsibility and support effective monitoring;
- Examples of systems for peer and self-reporting; and
- Other resources, such as occupational health and employee assistance programs, that can assist monitoring efforts.

The Importance of a Process or System

The recommendation above is supported by research from a variety of situations and settings about the general importance of having systems or processes in place to support positive action, including monitoring of potential problems among employees (Turner and Gelles 2003). Studies of organizations, such as those focused on fostering a productive organizational culture or understanding the dynamics of "high reliability" organizations where the costs of failure would be extremely high (e.g., air traffic control systems, nuclear power plants, the airline industry), also identify the importance of processes (Schulman et al. 2004; Weick and Sutcliffe 2001). All processes are not equal, as one would expect, and there may be significant challenges to creating those that can be both trusted by those involved to protect individuals and accepted by management as not posing a threat to their responsibilities and authority. The committee heard presentations about some types of processes in other sectors; Box 4-1 offers examples from the airline industry.

The literature on insider threats further argues that an organization should have the necessary processes in place before the problem occurs (Turner and Gelles 2003). An already existing process is much more likely to be effective

BOX 4-1
Examples of Screening and Peer Reporting Systems from the Airline Industry

Airline Pilots

Screening: After a conditional offer of employment, a number of checks: FBI fingerprint check/criminal history; National Driver Registry; previous employer; and Department of Transportation Drug and Alcohol Testing. Educational credentials may or may not be checked and references may or may not be checked.

Reporting: The Airline Pilots Association, the pilots union, operates a two-tier reporting system.

- The *Professional Standards Committee* (PSC) is a union committee that uses trained volunteers. The PSC operates through peer-to-peer interaction to deal with interpersonal conflict and stress-caused issues that peers believe mean a pilot should not be flying. It is reactive, not proactive, and its work is usually initiated by peers. The PSC is independent of airline management and never represents pilots to management. It operates under rules of strict confidentiality and keeps no written records. Discussions of cases are limited to those involved, and some have agreements with management denying access to information. The PSC may refer to outside mental health expert or to CIRP (see below). Since having such a committee lowers management costs, the process has the support of management.
- The *Critical Incident Response Program* (CIRP) team is a union committee that responds automatically to certain events. For example, the CIRP would send a team in the event of a crash or a major incident, comparable to events to which National Transportation Safety Board teams would respond. CIRP action may also be initiated by a pilot or peers or the PSC. The CIRP is staffed by highly trained pilot volunteers who go through many stages of training and tend to remain active with the committee for many years. The CIRP may also refer a pilot to an outside mental health professional.

Flight Attendants

Reporting: The flight attendants union also has a two-tier system.

- The main source of assistance is via *employee assistance programs* (EAPs). The EAPs handle referrals for interpersonal issues, performance deterioration, or unusual behavior. A referral may be self-generated or come from peers, a supervisor, or a union representative. The attendants' CIRP may also make a referral. The EAP may refer an attendant to an outside mental health professional. Training is conducted with managers, including in conflict resolution.
- In addition to the EAP, each crew base has a *Professional Standards Committee*, which deals with more diverse issues than the EAP.

SOURCE: Damos 2009.

than an ad hoc response, both for prevention and for responding when warning signs of imminent trouble appear. Identifying early warning signs will not necessarily reveal an insider before an incident occurs, but it can help identify individuals who might require assistance from trained professionals. Without this intervention, a particular individual may or may not resolve the situation on his or her own, but having measures in place to assist—rather than automatically exclude individuals showing signs of trouble—benefits everyone. It is important for those considering the creation of monitoring processes to note that one may find parts of the system are already extant within many organizations but serving other purposes; these systems can, therefore, be supplemented or adapted for this purpose.

Finally, a common message in National Academies reports on topics as disparate as medical error and assessment of U.S. democracy assistance programs (IOM 2000; NRC 2008c) is how important it is for an organization to be able to learn from mistakes and less successful endeavors as well as from triumphs. This is a broader organizational challenge than creating mechanisms to prevent insider threats, but the literature on "learning organizations" offers a range of models and lessons that provide some useful context for the specific problems addressed in this report (Schön 1973; Senge 1990).

Given how rare instances of attempted bioterrorism have been, the committee believes it would be helpful to develop case studies to explore examples of potentially relevant behavior (e.g., complacency, exploitation, theft of materials, scientific fraud) that have occurred specifically in the biosciences. Case studies already exist on some of these issues (e.g., Macrina 2005; NRC 2009c), so an important task is to identify and supplement relevant existing case studies, commission new ones, and examine them all comparatively to highlight lessons relating to personnel reliability and security.[15] As in evaluation efforts described in Chapter 5, these case studies will help move the policy discussions toward a better understanding of how to address the risk of the insider threat.

The Challenge to Management

As with screening, reducing the risk of an insider threat can be viewed as part of the larger set of challenges facing any manager. Successful programs to monitor and manage problems in the workplace unfortunately involve hard work and diligence. But there are reasons in addition to security to improve the quality of management in the workplace. In the case of BSAT, safety is clearly a primary reason, because that which improves safety generally will also enhance security. The changing environment in many laboratories, with greater emphasis on teamwork and larger groups of researchers, also makes management and

---

[15]See George and Bennett (2005) for a discussion of how such comparative case studies can be designed to be more methodologically rigorous.

mentoring important components of the job for any supervisor. Implementing changes to improve managing for security belongs within such existing systems. It is important to remember, however, that some actions taken to enhance security may not be relevant for what is done for ensuring safety. For example, will the "culture of trust" discussed below, which accepts peer reporting of potentially disqualifying behavior to ensure everyone's safety in the laboratory, necessarily extend to security, which introduces issues of criminal behavior and national security?

With this brief introduction, consideration is now given to research, evidence, and experience that can inform development of systems to improve personnel reliability at institutions working with BSAT materials through fostering active monitoring and management.

Fostering a Culture of Trust and Responsibility

A goal in any organization where safety is a central challenge should be to foster a culture where individuals watch out for each other and take responsibility for both their own performance and that of others. When this works well, the environment and culture reinforce a positive and inclusive ethic that promotes excellent performance. In turn, a balance of formal and informal processes will help to maintain the culture. Many of the components of a safety-oriented culture will serve security goals as well.

A successful culture of trust and responsibility relevant to personnel reliability requires the engagement of everyone in the laboratory. A key component is a climate inducing self- and peer-reporting and providing mechanisms for such reporting. On a cautionary note, understanding of the culture within a particular workplace or an organization before trying to use it to foster new practices is essential. Not taking culture into account can doom the effort to failure or inadequacy (Morgan 1997; Schein 2001).

As discussed in Chapter 1 and below, the culture of science already contains many of the elements conducive to fostering trust and responsibility. In addition, education and training provided to life scientists at different stages of their careers provide venues for the information the committee recommends. Fortunately, there is already movement in this direction for other reasons, upon which the Select Agent Program can build.

Essential Role of Education and Training

Good mentoring and training are important ways to develop a culture of responsibility, providing the essential foundation on which other elements of effective monitoring can be built. They should be viewed as necessary but not sufficient conditions, with continuing efforts by laboratory managers, researchers, and staff needed to sustain the culture and reinforce expectations of appropriate behavior. Training and educational experiences will have to be multifaceted to address the many interrelated issues in promoting the culture of trust and responsibility that Recommendation 1 seeks to instill. Although some training will need to be tailored to particular segments of the BSAT community, at least some discussions need to include everyone. It seems particularly important, for example, to foster discussions among and between the scientific and technical staff and those with responsibility for security so that common understanding can be built.

Leadership Development

Incorporating engagement as a critical factor in managing a safe and secure workforce should be part of leadership development, but the requirement for engaged management may not come naturally to laboratory managers and officials. Most scientific laboratory managers attain their position by intellectual achievement. The qualities that lead to success as an outstanding researcher do not necessarily relate to management skills. Moreover, many who are promoted to supervisory positions in laboratories are not provided opportunities for training in management. Nevertheless, there are many good laboratory managers who do provide engagement and oversight.

Since the principal investigator is the most likely individual to interact regularly with a broad cross-section of research and technical staff, he or she will need particular support in the form of resources to acquire the skills needed for effective engagement and monitoring. The diversity of facilities carrying out BSAT research makes it difficult to offer generalizations about approaches to this kind of leadership development that would apply nationally. Such resources may be more readily available in federal laboratories or private and commercial entities, where management training is more often provided and encouraged, than in academic environments, where managers maintain a greater degree of independence. Moreover, the opportunity to develop and/or sharpen management skills seems less likely to be seen as important for an academic scientific career than for a career in private or government environments.

Education to Raise Awareness and Foster Responsibility

Building a culture of trust and responsibility to reduce the risk that BSAT materials might be stolen for use by terrorists or used in acts of sabotage in the laboratory can draw upon longstanding traditions in the life sciences as well as more recent efforts focused on security risk. The iconic example of the life sciences' exercising responsibility is its response in the early 1970s to concerns about potential safety risks arising in the then newly developing field of recombinant DNA research. The 1975 Asilomar Conference on Recombinant DNA brought scientists together to discuss risks of manipulating DNA from different species. The results of the meeting led to the National Institutes of Health's (NIH) issuing its *Guidelines for Research Involving Recombinant DNA Molecules* and creation of a process for reviewing proposed experiments that continues today.[16] The Human Genome Project created an ethical, legal, and social implications program to explore how advances in genetics intended to improve human health could proceed without undermining other dimensions of human well-being.[17]

More recently, concerns have been raised about the so-called "dual use dilemma" of the life sciences, in which results of research intended for beneficial purposes, such as therapies against infectious diseases, might be misused for biological weapons or bioterrorism. This has led to calls for educating the life sciences community about its responsibility to reduce such risks. Dual use research is a broader concept than BSAT, but it is reasonable to assume that much of the research conducted under the Select Agent Program could potentially be considered dual use. A series of NRC reports has endorsed education on dual use issues (NRC 2004ab, 2006, 2007b, 2009ab). The NSABB has proposed that all federally funded researchers in the life sciences receive training about dual use issues (NSABB 2007) and, at the time of this report, the proposal is under review by an interagency working group. The American Association for the Advancement of Science (AAAS) and the Federation of American Societies for Experimental Biology (FASEB) have also recommended training programs, although both stop short of recommending that such programs be mandatory (AAAS 2008; FASEB 2009). This suggests that education for BSAT researchers might be able to draw on and fit within at least some of these initiatives, especially if the NSABB's recommendation of mandatory training is adopted.

In most cases, recommended training on dual use issues is viewed as becoming part of other, broader training for life scientists on responsible conduct, rather than as standalone activities. In the United States, there are three types of existing education to which the kind of training envisioned by the committee naturally might be added.

- Biosafety training has not traditionally included security issues, but there is evidence that some training programs have added discussions and modules (AAAS 2009; the appendix includes a list of training programs). This might be the venue best able to reach the full range of laboratory technical and research staff, as well as those outside academia.
- NIH mandates training in the responsible conduct of research (RCR)

---

[16]The current version of the Guidelines is available at <http://oba.od.nih.gov/rdna/nih_guidelines_oba.html>. The first revisions to the scope of the Guidelines are currently being made to reflect the implications of the new field of synthetic genomics.

[17]See <http://www.ornl.gov/sci/techresources/Human_Genome/research/elsi.shtml> for more information.

for those who are supported by its training grants (NRC 2009c). RCR training is frequently cited as the most promising U.S. venue for dual use education; although its scope would have to be expanded beyond the current focus on research integrity and beyond those supported by its training grants to reach a much broader segment of life scientists, there are signs the RCR community is interested in the opportunities dual use education offers (AAAS 2008).

- Bioethics training, which largely reaches those in biomedical research including BSAT researchers, offers another potential venue, and, again, there are signs of interest from some in that community in taking on the additional issues (AAAS 2008).

It was not within the committee's charge to offer highly specific recommendations on how best to undertake the education and training needed to foster the culture of trust and responsibility that is recommended. The committee believes, however, that whatever venue is chosen—and all of them might be appropriate for particular contexts in order to reach the range of BSAT research entities—educational materials will need to be developed and resources provided to support and sustain implementation. Box 4-2 offers two

---

BOX 4-2
Sample Educational Materials for Considering
Dual Use Research Issues

The Federation of American Scientists' *Case Studies in Dual-use Biological Research* illustrate the "dual use" potential of actual life science research. The case studies provide a historical background on bioterrorism and bioweapons and the current laws, regulations, and treaties that apply to biodefense research. They include interviews with researchers as well as the primary scientific research papers and discussion questions meant to raise awareness about the importance of responsible biological research. The case studies are available at <http://www.fas.org/programs/ssp/bio/educationportal.html>.

The Policy, Ethics and Law Core of the NIH-funded Southeast Regional Center of Excellence in Biodefense has developed an online module to assist those involved with the biological sciences to better understand the "dual use" dilemma of some life science research. This module is intended for graduate students and postdoctoral scholars, faculty members, and laboratory technicians involved in biological research in microbiology, molecular genetics, immunology, pathology, and other fields related to emerging infectious disease and biodefense. The module consists of an approximately 20-minute online presentation followed by a brief assessment and has been used by more than 600 people. The module is available at <http://sercebtraining.duhs.duke.edu/>.

examples of current online resources to illustrate some of the types of materials that would be needed. The report of a workshop on ethics education held in 2008 by the National Academy of Engineering's (NAE's) Center for Engineering, Ethics and Society (NAE 2009) offers an introduction to some of the current thinking about the components of effective ethics education in science and engineering. The NSABB's strategic plan for outreach and education on dual use issues (NSABB 2008) and the AAAS workshop on dual use education (AAAS 2008) offer ideas focused more directly on BSAT-related issues. Here, the agencies overseeing and supporting BSAT research and the professional societies, separately and in collaboration, can play a major role in supporting and disseminating materials and sharing successful practices.

Current Training by Registered BSAT Entities

The Select Agent Program requires all registered entities to provide training in biosafety and security before individuals can enter areas where select agents and toxins are handled or stored (7 CFR 331.15(a) and 9 CFR 121.15(a)). The training "must address the particular needs of the individual, the work they will do, and the risks posed by the select agents or toxins." Annual refresher training is required, and an entity's training program is included in inspections conducted by the Centers for Disease Control and Prevention (CDC) and the Animal and Plant Health Inspection Service (APHIS). Current training programs are primarily technical, focusing on biocontainment and biosafety practices and the details of a facility's security plan.

Required training offers an opportunity to reach all participants in BSAT research with at least some essential messages that will promote personnel reliability. Support for this type of training was strongly endorsed during one of the committee's site visits. The committee believes that, without substantially increasing the time entailed in security training, a module focused on the risk of an insider threat could be added. At a minimum such a module could include the likelihood of warning signs and examples of what they might be, the expectation for peer and self-reporting, and the resources available to make a report. CDC and APHIS could work with federal security agencies or with outside experts to develop relevant materials for use by entities or provide resources that entities could use to develop their own. Discussions about individual responsibility and updates on available resources could be part of the required refresher training.

Systems for Peer and Self-Reporting

Specific examples of programs already exist in many laboratory settings to assist with some of the aspects of monitoring behavior as part of safety that can support monitoring for security as well. When warning signs appear, peers and colleagues are most likely to be in a position to notice them. Part of the culture of trust and responsibility includes individuals' feeling encouraged to report on themselves and others if they find signs of trouble or feel that an individual poses a safety or security risk. To enable coming forward, it is important to provide reporting mechanisms that individuals trust. Management plays an essential role and has important responsibilities. It is management's responsibility, for example, to provide or permit mechanisms for people to self-report problems and relay concerns about others via a safe mechanism (e.g., ombuds offices, hotlines, and/or confidential reporting systems). Management may also provide mechanisms for individuals to obtain help in dealing with concerns proactively via employee assistance programs (EAPs). Although often focused on safety concerns, these processes can serve security as well.

In creating "safe" reporting mechanisms, it is important to be sensitive to management's need for information and its ultimate responsibility for whatever happens. In some cases, including the BSAT program, there may be legal requirements, including potential civil or criminal penalties for noncompliance, as part of managers' responsibilities to assure security. Encouraging reporting can be difficult even if most of those working in a facility believe that they can trust their managers. Where managers are considered "part of the problem," the difficulties of creating effective reporting mechanisms multiply. Simply requiring such reporting is not an answer if the basic culture of trust is absent. In fact, imposing reporting requirements in the wake of an incident may have negative consequences, unless those affected believe it is part of a positive change and is not punitive or palliative.

Reporting Systems: The Ombuds   Many organizations that made presentations during the public consultations for the NSABB report and the EO Working Group described reporting systems for identifying problems. In addition, the committee heard a presentation from two experienced ombuds at its first meeting, who reported their research on why people do or do not report "inappropriate" behavior:

> Most people consciously or intuitively consider the context when they perceive behavior that they think is wrong. They may consider the rules—and also the actual norms—of their *organization*, about acting on the spot or "coming forward." They may review their own and their colleagues' perceptions of the *local supervisor*. They may, consciously or intuitively, evaluate their *complaint system and its options*, in terms of safety, accessibility and credibility. Recent events may also affect peoples' actions.

> Personal factors include how people understand the issues at hand, their personal preferences, gender and cultural traditions, and their perceived power or lack of power. People also may behave differently depending on their role in the situation—as an injured party, a perpetrator, supervisor, senior officer, peer or "bystander." (Rowe et al. 2009:10)

Although not directed at the problem of preventing an insider threat of bioterrorism, the findings are informative for thinking about the design of systems for reporting and self-reporting. The research cited above concluded that: *"There is no single policy that will make an organization seem trustworthy and no single procedure or practice that will guarantee that people will overcome all the barriers to coming forward.* A well-publicized commitment to fairness and to procedural justice may be a good beginning" (Rowe et al. 2009:24; italics in original). The findings of an extensive review of the literature on reporting systems identified five "core" characteristics of the most effective ones, which are summarized in Box 4-3.

Occupational Health Programs A monitoring program intended to identify problems before they occur may take advantage of programs already in existence for other purposes. In situations where the health and safety of workers is a

---

BOX 4-3
Core Characteristics of Effective Reporting Systems

Elegance—simple to understand, apply to a broad range of issues, and use an effective diagnostic framework. Those who manage the system should be able to respond definitively to the issues raised.

Accessibility—easy to use, with information about how to report or file a complaint widely advertised and readily comprehensible.

Correctness—(1) relevant input about the problem can be reported, (2) the organization can investigate and call for more information if needed, (3) a system exists for classifying and coding information in order to determine the nature of the problem, (4) employees can appeal lower-level decisions, and (5) both procedures and outcomes make good sense to most employees.

Responsiveness—at the most basic level, responsive systems let individuals know that their input has been received. Responsive systems provide timely responses, are backed by management commitment, are designed to fit an organization's culture, provide tangible results, involve participants in the decision-making process, and give those who manage the system sufficient clout to ensure that it works effectively.

Nonpunitiveness—essential if employees are to trust the system. Individuals must be able to present problems, identify concerns, and challenge the organization in such a way that they are not punished for providing this input, even if the issues raised are sensitive and highly politicized. If the input concerns wrongdoing or malfeasance, the individual's identity must be protected so that direct or indirect retribution cannot occur. Employees as well as managers must be protected.

SOURCE: Sheppard et al. 1992.

major and continuing concern, institutions may have ongoing relationships with occupational health physicians.[18] The current edition of *Biosafety in Microbiological and Biomedical Laboratories*, which is directly relevant to BSAT research, discusses the need for an occupational health program, although it does not specifically include mention of resources to address mental or emotional health (CDC/NIH 2007). Several representatives of BSAT research facilities who spoke at the public consultations for the NSABB and EO Working Group described these types of arrangements. In some cases, physicians may be responsible for periodically reviewing and certifying the continued fitness of workers, including their mental health (perhaps in consultation with mental health specialists as needed). In others, occupational health professionals are on call to provide assistance to management or employees. When these arrangements work well, they provide a source of assurance that those working in BSAT laboratories are physically and mentally fit to be carrying out their research.

Employee Assistance Programs   In addition to, or in some settings instead of, occupational health programs, employee assistance programs may play a role. EAPs are benefit programs offered by employers, usually at no cost to employees and as an adjunct to health insurance plans. Very generally, EAPs are intended to assist employees in addressing personal problems that might negatively affect their performance at work (e.g., substance abuse, major life event, financial or legal issues, family relations, workplace relations, etc.). Some degree of assessment, short-term counseling, and referral services are typical components of an EAP. Many employers contract with an outside firm to provide EAP services, since the range and variety of issues that may arise are likely to be beyond the expertise of a normal human resources office. Most EAPs have toll-free numbers that provide round-the-clock access, which is also beyond the capacity of most institutions. In addition to providing confidential resources that employees may seek on their own, in some cases employers may refer employees for performance-related issues. EAPs have the advantage of relieving a manager of the expectation that he or she will be able to diagnose specific problems; instead, the manager's role is to identify declining work performance and then refer the employee to the EAP.

Summing Up   Having occupational health specialists available and active in monitoring laboratory personnel could provide genuine assistance in monitoring for insider threats, at least for the type of behavior that is most likely to

---

[18]This section does not address legal requirements that might be imposed by the Occupational Health and Safety Administration (OSHA). Note that the Select Agent Program includes regulations and guidance from OSHA as one of the set of suggested standards for ensuring biosafety requirements of the program (see <http://www.osha.gov/pls/oshaweb/owadisp.show_document?p_id=3359&p_table=OSHACT>).

be detected. An EAP could support such efforts, and the fact that these types of programs already exist for other purposes has a significant cost advantage. There is the possibility that someone suffering from personal stress that might lead him or her to undertake, or be an accomplice to, terrorism might seek help and in doing so provide an alert.[19] If employees believe that seeking help, which might include taking themselves out of the laboratory for a period of time, will not have a significant impact on their careers, then the existence of such programs could have substantial benefit in avoiding larger problems. It is important to note that standard practice in industry is to return people to their safety-sensitive jobs after treatment. This is consistent with the Americans with Disabilities Act and also assures affected employees that partaking in interventions such as through an ombuds office, occupational health program, or EAP does not have "career-ending" consequences.

Implementation   Because the committee did not conduct a detailed review and assessment of current and potential programs for monitoring, this report can only raise a number of issues that need to be considered in the development and implementation of such programs. The committee notes that these programs could contribute substantially to a safe and secure research environment, although we emphasize that no system can provide complete insurance against the risk of an insider threat.

One of the most difficult issues involved in creating a monitoring program is ensuring that peer or self-reporting can be done in a manner that is not damaging or "career-ending." This applies across the board for all kinds of behavior but becomes increasingly difficult as the behavior in question moves toward instances where someone's conduct is potentially negligent or even criminal. In such instances, reporting in a system built for safety rather than security becomes problematic. What will be the consequences for a person who reports on him- or herself? Is it more important to learn about the behavior, correct any damage, and perhaps find ways to avoid similar behavior in the future or to be sure that there are consequences for inappropriate actions? When are "zero tolerance" policies productive by establishing clear rules and when are they counterproductive by making people feel they dare not report even unintentional lapses? What happens to a person who "blows the whistle" on a colleague? Which disincentives, such as a fear of being sued, might keep managers from acting on warning signs, which in addition to any security risk, could undermine the integrity of the reporting system?

Moving beyond the particular facility or organization, what is the role of those charged with regulating BSAT entities? How do hotlines or other

---

[19]Privacy concerns as well as norms of confidentiality, reinforced by local, state, or national regulations and laws, might hinder reporting these signs to institutional management or authorities, unless protocols were already in place.

reporting mechanisms maintained by regulatory agencies fit into the picture? What is an appropriate mix, if any, between "compliance assistance," which might include permitting entities to report violations provided culpability is acknowledged or a plan is in place to prevent future transgressions, and "enforcement"?

One message from many of the presentations and the committee's own discussions was the importance of keeping oversight for reporting systems at the local level to the extent possible. Some of this may reflect the natural reaction of the regulated or potentially regulated to yet another requirement from higher authorities. But there was also a strong sense that many if not most of the problems identified by a reporting system would be most readily and effectively dealt with at the local level. And there is already considerable information flowing upward: the current incident and theft and loss reporting requirements, for example, already provide CDC and APHIS with information about operations in BSAT facilities.

For a number of reasons, including increasing the chances of an appropriate and effective response in the event of an incident, it may also be important to use opportunities provided by internal institutional reporting systems to establish constructive relationships with local law enforcement and FBI officials. Anecdotal evidence suggests that building these relationships is challenging, yet the effort involved may be worth it if it contributes to overcoming some of the concerns and "culture clash" found in a 2008 survey by the Federation of American Scientists and the FBI on how scientists view law enforcement (Hafer et al. 2008).[20]

Summing Up

One of the most important recommendations in this report is to foster a culture of trust and responsibility in the laboratory and undertake education and training to support it. The BSAT research community—and the life sciences community more broadly—has a responsibility to help reduce the risk that the results of the knowledge, tools, and techniques developed for beneficial purposes are not misused. Given that no personnel screening process can be

---

[20]"The attitudes of scientists toward law enforcement personnel are not vastly different from those of the general public…. However, a larger percentage of scientists indicated cooler feelings towards the FBI than the general public, suggesting that these reservations are particular to the scientific community and require specific solutions with the scientific community in mind. The results show that scientists hold more favorable feelings towards state and local law enforcement than federal law enforcement. However, when confronted with specific issues or concerns, the responses reveal no significant distinction between interacting with the FBI or with law enforcement in general" (Hafer et al. 2008). The survey, conducted among AAAS members, suffered from a low response rate (just above 13 percent) but those who responded indicated a clear preference for local control.

expected to predict the behavior of employees in all contexts at all times, active management, monitoring, and support for those working in BSAT laboratories are key components of a comprehensive approach that builds trust and ultimately leads to safer and more secure BSAT research. Such programs are already in place in some of the laboratories carrying out BSAT research, but to be fully effective this type of program needs to be transformed into standard practice throughout the Select Agent Program.

5

Managing BSAT Research and the Select Agent Program

INTRODUCTION

Research with biological select agents and toxins (BSAT) requires compliance with a number of policies and procedures designed to maintain the security of these materials. The issues related to personnel training, screening, and monitoring are discussed in Chapter 4. This chapter addresses other aspects of the program identified by the committee as important, along with the committee's conclusions and recommendations.

The issues addressed here relate to stakeholder engagement in the Select Agent Program, the list of select agents and toxins, the need for accountability, security based on risk assessment, the role of evaluation in moving forward, improvement of the laboratory inspection system and inspector training, and mechanisms needed to fund security and compliance.

FACILITATING STAKEHOLDER INPUT BY FORMATION OF A BSAT ADVISORY COMMITTEE

One of the frequent themes that emerged from the public consultations held by the Executive Order (EO) Working Group on Strengthening the Biosecurity of the United States and the National Science Advisory Board for Biosecurity (NSABB) and in the committee's own public sessions and site visits was the need for increased and more systematic communication among those agencies funding BSAT research, those agencies administering the Select Agent Program, and those entities conducting BSAT research. The Centers for Disease Control and Prevention (CDC) and the Animal and Plant Health Inspection Service (APHIS), through their training programs and outreach at professional meetings, are already providing information and guidance to the regulated community. The creation of the National Select Agent Registry as a single point of contact for agents regulated by both CDC and APHIS has been almost universally applauded for simplifying the regulatory environment and providing coordinated guidance. But because BSAT research is carried out and supported by several federal agencies, not just the Department of Health and Human Services (HHS) and the U.S. Department of Agriculture (USDA), the committee believes a more formal structure is needed to engage the community of stakeholders in the operation of the program.

To provide a locus for both feedback from the research community and discussion of issues of common interest, the committee recommends the establishment of a Biological Select Agents and Toxins Advisory Committee (BSATAC), whose membership would be drawn from the BSAT research community. This committee would provide an ongoing conduit for discussion of the implementation of the select agent regulations and would be made up of microbiologists and other infectious disease researchers (including select agent researchers) at various career stages, responsible officials, and those with experience in biosecurity, animal care and use, compliance, biosafety, operation of BSAT facilities as well as viewpoints from the public health, risk assessment, and legal communities. Representatives from the federal agencies with a responsibility for funding, conducting, or overseeing select agent research would serve in an ex officio capacity.[1] The committee would operate under the provisions of the Federal Advisory Committee Act, which mandates public meetings, adequate advance notice, and public accessibility of records—all of which would serve to broaden the reach of the BSATAC.

Such an advisory committee should have several specific responsibilities. Based on results of the consultations and site visits, it is clear that an important responsibility for the BSATAC would be the promulgation of guidance on the implementation of the Select Agent Program. A survey of the principal investigators (PIs) and co-PIs of the National Institutes of Health (NIH) Regional Centers of Excellence (RCEs) for Biodefense and Emerging Infectious Diseases found that more than 90 percent believed that select agents should be regulated, but also that a majority were concerned about whether or not they were in compliance with the rules and that they lacked a source to go to with their questions (Sutton 2009). In addition to providing formal guidance, the Advisory Committee would facilitate exchange of information about the program—such as aggregate data on laboratory-acquired infections—and promote the sharing of

---

[1] Among the agencies it would be appropriate to have represented are the Office of Science and Technology Policy; the National Security Council/Homeland Security Council; the Office of Management and Budget; the Department of Health and Human Services; the Department of Agriculture; the Department of Justice; the Department of Homeland Security; the Department of Defense; the Department of Energy; the Department of Transportation; the Department of Commerce; the Department of State; the Environmental Protection Agency; the National Science Foundation; and the Office of the Director of National Intelligence.

successful practices across institutions and sectors. Both of these actions would substantially assist in the implementation of the program. At present, they do not occur on a regular basis, to the detriment of efficiency and uniform compliance. To carry out these responsibilities, the Advisory Committee would convene regular public meetings of key constituency groups. All of these actions would contribute to another important potential Advisory Committee responsibility: promoting harmonization of regulatory policies and practices.

In addition to these responsibilities, the Advisory Committee should fulfill two other important functions. Below, we recommend that the Select Agent Program carry out regular evaluation of the relative security benefits it provides and of their consequences, both intended and unintended, that their implementation is having on BSAT research. We believe these evaluations are necessary, whether or not the recommended Advisory Committee is created, but it would be a natural body to provide oversight for the conduct of such evaluations and for determining how the results are used by the Select Agent Program.

We also believe the Advisory Committee should provide advice on the composition and/or stratification of the list of select agents and toxins, as discussed below. The BSATAC would not be a substitute for the rulemaking required by statute, but it could be a source of important input to decision-making and provide a mechanism for ongoing community engagement.

> **RECOMMENDATION 2:** To provide continued engagement of stakeholders in oversight of the Select Agent Program, a Biological Select Agents and Toxins Advisory Committee (BSATAC) should be established. The members, who should be drawn from academic/research institutions and the private sector, should include microbiologists and other infectious disease researchers (including select agent researchers), directors of BSAT laboratories, and those with experience in biosecurity, animal care and use, compliance, biosafety, and operations. Representatives from the federal agencies with a responsibility for funding, conducting, or overseeing select agent research would serve in an ex officio capacity. Among the responsibilities of this advisory committee should be the following:
>
> - Promulgate guidance on the implementation of the Select Agent Program;
> - Facilitate exchange of information across institutions and sectors;
> - Promote sharing of successful practices across institutions and sectors;
> - Provide oversight for evaluation of the Select Agent Program;
> - Provide advice on composition/stratification of the list of select agents and toxins;
> - Convene regular meetings of key constituency groups; and
> - Promote harmonization of regulatory policies and practices.

This committee is not recommending a particular site for the Advisory Committee within the federal government, although there appear to be two natural options. One choice would be to make BSATAC a joint enterprise for the Secretaries of HHS and USDA, since these two departments have the formal statutory authority for the Select Agent Program. The secretariat might then reside within one or the other agency, perhaps with the Select Agent Program office in CDC. This would have the advantage of locating the Advisory Committee with the program to which it is providing advice, but might overemphasize the role of the regulatory agencies with responsibility for the program and make it more difficult to consider the other federal agencies that are essential to effective functioning of the program.

Another option would be to make the committee a project of the Office of Science and Technology Policy (OSTP) within the Executive Office of the President, perhaps with a formal link to the National Security Council. In this case, the Director of OSTP would appoint the membership of the committee. This would give the Advisory Committee—and BSAT research more generally—greater visibility and allow OSTP to perform a coordinating function for science and technology policy that is one of its normal tasks.

Wherever the Advisory Committee is located, it will need a small secretariat and professional and administrative staff to support its work as well as dedicated funding. Creating the Advisory Committee without providing it with the necessary resources to carry out its responsibilities, or forcing its host agency to absorb the costs, would not be appropriate and thereby substantially reduce its chances of success.

A number of similar committees exist throughout the government to provide a mechanism for the formal engagement of key external stakeholders. For example, the Secretary's Advisory Committee on Human Research Protections (SACHRP) provides expert advice and recommendations to the HHS Secretary and Assistant Secretary on issues related to human research subjects and reviews selected work and activities within HHS.[2] Similar to the proposal for BSATAC, SACHRP is composed of experts from outside the government along with nonvoting ex officio members from the relevant federal agencies.

STRATIFICATION OF THE LIST OF SELECT AGENTS AND TOXINS

The current list of select agents and toxins contains more than 80 entries.[3] The list represents a diversity of pathogenic microorganisms and biological toxins with a range of potential for use as biothreat agents. This prompted

---

[2]See <http://www.hhs.gov/ohrp/sachrp/> for more information about SACHRP.

[3]In addition, there are pending proposals from the Department of Health and Human Services to add the SARS-associated coronavirus and Chapare virus to the list (HHS 2009ab).

the committee to ask if a single list of select agents and toxins, all of which are subject to the same security procedures, represents an optimal solution.

A very few items on this list have actually been used as a biothreat agent, and there is legitimate reason to be concerned about these items (see reviews by Carus 2001 and Wheelis et al. 2006). There is also reason to have security concerns about particularly deadly pathogens that may be genetically engineered to become more dangerous or resistant to treatment or those that can be effectively disseminated. But do all of the materials on the list prompt the same level of concern and risk? Does including a microorganism or toxin that poses a minimal security risk justify the potential lost capacity for research on that agent? Is a single undifferentiated list of select agents and toxins the safest and most effective way to ensure the security of the American public?

In addition to the biosafety classifications discussed in Chapter 2, several stratifications of the select agent list already exist (see Box 5-1). For example, CDC has defined Category A, B, and C bioterrorism agents, with Category A agents explicitly recognized as "organisms that pose a risk to national security." The NIH *Guidelines for Research Involving Recombinant DNA Molecules* define four risk groups to catalogue biohazardous agents, including select agents. The Department of Homeland Security (DHS) conducts a bioterrorism risk assessment (BTRA) that encompasses a number of factors including the agents, routes of acquisition, methods of production and weaponization, targets, modes of dissemination, public health consequences and economic cost (Pesenti 2009).[4] A Blue Ribbon Panel convened by OSTP prioritized a list of animal pathogens of the greatest concern (Kelly et al. 2004). In addition, other national priorities, such as the list of agents against which we develop countermeasures or stockpile vaccines provides some indication of the agents that prompt the most concern. Finally, a graded approach to security has been ongoing for some time in work with nuclear materials, so there is precedent for implementing different security procedures based on risk.[5]

The committee concluded that the present all-encompassing model for the list of select agents and toxins does not address appropriately the range of risks and vulnerabilities presented by these agents. Moreover, a list of more than 80 agents of varying risk dilutes attention from those that pose the greatest degree of concern, which may, in the process, render the nation less secure. It would be more effective to focus the highest scrutiny on those agents that are, indeed, of greatest concern and on those facilities with the equipment that enables weaponizing biological agents—and to offer a graded series of security procedures and policies for agents that pose less risk. For these reasons, the committee recommends a reconsideration of the purpose and composition of

---

[4]See NRC (2008a) for a recent review of the BTRA.

[5]Box 5-3 provides additional comparisons between nuclear and biological materials, including the security implications.

BOX 5-1
Selected Existing Classifications of Biological Agents

Centers for Disease Control and Prevention (reproduced from <http://www.bt.cdc. gov/agent/agentlist-category.asp>)

Category A: High-priority agents include organisms that pose a threat to national security because they:

- can be easily disseminated or transmitted from person to person;
- result in high mortality rates and have the potential for major public health impact;
- might cause public panic and social disruption; and
- require special action for public health preparedness.

Category A agents/disease include anthrax (*Bacillus anthracis*), botulism (*Clostridium botulinim* toxin), plague (*Yersinia pestis*), smallpox (variola major), tularemia (*Francisella tularensis*), and viral hemorrhagic fevers (filoviruses [e.g., Ebola, Marburg] and arenaviruses [e.g., Lassa, Machupo]).

Category B: Second highest priority agents include those that:

- are moderately easy to disseminate;
- result in moderate morbidity rates and low mortality rates; and
- require specific enhancements of CDC's diagnostic capacity and enhanced disease surveillance.

Category B agents/diseases include brucellosis (*Brucella* species), epsilon toxin of *Clostridium perfringens*, food safety threats (e.g., *Salmonella* species, *Escherichia coli* O157:H7, *Shigella*), glanders (*Burkholderia mallei*), melioidosis (*Burkholderia pseudomallei*), Psittacosis (*Chlamydia psittaci*), Q fever (*Coxiella burnetii*), ricin toxin

---

the list of select agents and toxins to reflect actual security concerns that merit inclusion on the list.

Although consideration of which specific agents and toxins should be on such a list is beyond the charge of the committee, the committee believes that stratification of the list of select agents and toxins is both warranted and necessary. Stratification should be consistent with the original purpose of creating the list, namely to catalogue those agents that pose a risk for use as a significant biothreat agent. If the purpose of the Select Agent Program were to protect the public from any infectious organism representing a threat to public health, then such a select agent list would be both unwieldy and highly disruptive to biomedical research—in fact, all infectious disease research would then be subject to inclusion on the list. If the purpose of the Select Agent Program is to secure and protect the public against microorganisms and toxins that might be used as biothreat agents where the consequences cannot be easily managed,

from *Ricinus communis* (castor beans), staphylococcal enterotoxin B, typhis fever (*Rickettsia prowazekii*), viral encephalitis (alphaviruses [e.g., Venezuelan equine encephalitis, eastern equine encephalitis, western equine encephalitis]), water safety threats (e.g., *Vibrio cholerae, Cryptosporidium parvum*).

Category C: Third highest priority agents include emerging pathogens that could be engineered for mass dissemination in the future because of:

* availability;
* ease of production and dissemination;
* potential for high morbidity and mortality rates and major health impact.

Category C agents include emerging infectious diseases such as Nipah virus and hantavirus.

National Institutes of Health

NIH Guidelines (available at <http://oba.od.nih.gov/oba/rac/guidelines_02/APPENDIX _B.htm>)

* Risk Group 1 (RG1): Agents that are not associated with disease in healthy adult humans.
* RG2: Agents that are associated with human disease which is rarely serious and for which preventive or therapeutic interventions are *often* available.
* RG3: Agents that are associated with serious or lethal human disease for which preventive or therapeutic interventions *may* be available (high individual risk but low community risk).
* RG4: Agents that are likely to cause serious or lethal human disease for which preventive or therapeutic interventions are *not usually* available (high individual risk and high community risk).

then the list is too long and should not include microorganisms with little or no potential for use as a biothreat agent or those whose impact can be effectively managed in other ways.

A team composed of subject matter experts from within the federal government—the Intragovernmental Select Agents and Toxins Technical Advisory Committee (ISATTAC)—currently advises the Select Agent Program on proposals to add or delete agents or toxins from the list.[6] At present there is

---

[6]The members of the ISATTAC are federal government employees from CDC, APHIS, NIH, the Food and Drug Administration, USDA/Agricultural Research Service (ARS), USDA/CVB (Center for Veterinary Biologics), and the Department of Defense (DOD). The group's three functions are to: (1) review requests for the exclusion of attenuated strains, (2) review requests to conduct restricted experiments, and (3) review requests for addition or deletion of agents or toxins to the list of select agents and toxins (Select Agent Program website: <http://www.selectagents. gov/FAQ_General.html#sec3q1>).

no mechanism beyond the formal rulemaking process through *Federal Register* notices for the BSAT research community and the broader community of infectious disease researchers to provide suggestions, comments, or advice to ISATTAC or to the Select Agent Program.

We believe that it is important to develop mechanisms for adding or removing agents from the list without unwarranted delay, to ensure that the list remains reflective of legitimate concerns. A procedure is needed to assess the threat risk posed by a biological agent that would initiate a formal process to add it to the list—or, equally important, to determine that an earlier estimation of threat has diminished and an agent should be taken off the list. For example, development of an effective treatment or vaccine would diminish the value of a particular pathogen as a biothreat agent, thereby downgrading its risk profile. Critical in consideration of adding or removing an agent from the list is the inclusion of significant information and input from external stakeholders, beyond the usual formal commenting process to government officials.

> **RECOMMENDATION 3: The list of select agents and toxins should be stratified in risk groups according to the potential use of the material as a biothreat agent, with regulatory requirements and procedures calibrated against such stratification. Importantly, mechanisms for timely inclusion or removal of an agent or toxin from the list are necessary and should be developed.**

As described in Recommendation 2 above, an external group of stakeholders—the Biological Select Agents and Toxins Advisory Committee—should be charged with advising on the composition of the list and facilitating wider engagement of the research and security communities. This Advisory Committee should also advise the Select Agent Program on the implications that stratification of the list of select agents and toxins has on implementation of personnel screening, physical security requirements, and other procedures.

It should be noted that addition of an agent or toxin to the list will have a significant impact on the conduct of research on that agent or toxin because the research will go from no special security requirements to the full complement of the select agent regulations. In effect, long-standing research programs will immediately be at risk if the institutions where the research is conducted are unable to take on the additional responsibilities for select agent research, if secure laboratory facilities cannot be obtained, or if researchers are unable to be cleared through the Security Risk Assessment (SRA) screening process. Even with a time for phasing in of implementation, how will the security enhancements be funded? What will happen if a researcher does not pass the SRA because of a past offense or other flag in the criminal, immigration, and terrorist database screen? What will be the impact of that exclusion on the researcher's career and on the progress of science?

Similarly, removal of a select agent or toxin from the list will immediately decrease the security requirements and restrictions. What will happen with unnecessary and excessively secure laboratory capacity and an expensive security apparatus that may no longer be needed?

ACCOUNTING FOR MATERIALS

It is prudent and appropriate for entities with the responsibility for BSAT laboratories to know what types of select agents and toxins are present in their facilities. In addition to maintaining records of materials in a facility for security purposes, such listings serve an important safety function in detailing materials of concern for laboratory personnel, as well as for first responders in emergencies.

Several sections of the select agent regulations discuss inventory control and recordkeeping associated with select agents and toxins. The section on records contains the most detailed guidance and is included in Box 5-2. In addition, the Select Agent Program provides further guidance and examples of the type of record-keeping that may be expected:

> **Inventory Control:** Each entity is required to keep a current and up-to-date inventory. How that inventory is conducted and maintained must be documented in the entity's security plan and must be consistent with the requirements found in Section 17 [see Box 5-2]. The select agents and toxins in the entity inventory must be labeled and identified in a way that leaves no question that what is in stock is accurately reflected in the inventory records.... All inventory records must be safeguarded to prevent alterations and be retained for 3 years.... (CDC/APHIS 2007)

Entities or individuals required to register under the Select Agent Program must develop and implement a written security plan that describes procedures for inventory control, the reporting of loss or theft of select agents and toxins, or the alteration of inventory records, among other elements. The security regulations also require that individuals with clearance to work with select agents are mandated to report the loss or theft of such agents or any sign that inventory or use records have been altered or otherwise compromised.

These regulations provide highly specific guidance with respect to information to be collected. While the committee believes that it is useful and important to know which agents are present and where they are located, we question the value of measuring the quantity for living microorganisms, except for the amounts when acquired by a facility or transferred out to another facility.

BOX 5-2
Select Agent Regulations—Records Section (42 CFR 73.17)[a]

§ 73.17 Records.
(a) An individual or entity required to register under this part must maintain complete records related to the activities covered by this part. Such records must include:

(1) Accurate, current inventory for each select agent (including viral genetic elements, recombinant nucleic acids, and recombinant organisms) held in long-term storage (placement in a system designed to ensure viability for future use, such as in a freezer of lyophilized materials), including:

    (i)   The name and characteristics (e.g., strain designation, GenBank Accession number, etc.),

    (ii)   The quantity acquired from another individual or entity (e.g., containers, vials, tubes, etc.), date of acquisition, and the source,

    (iii) Where stored (e.g., building, room, and freezer),

    (iv) When moved from storage and by whom and when returned to storage and by whom,

    (v)   The select agent used and purpose of use,

    (vi) Records created under § 73.16 and 9 CFR 121.16 (Transfers),

    (vii) For intra-entity transfers (sender and the recipient are covered by the same certificate of registration), the select agent, the quantity transferred, the date of transfer, the sender, and the recipient, and

    (viii) Records created under § 73.19 and 9 CFR part 121.19 (Notification of theft, loss, or release),

(2) Accurate, current inventory for each toxin held, including:

    (i)   The name and characteristics,

    (ii)   The quantity acquired from another individual or entity (e.g., containers, vials, tubes, etc.), date of acquisition, and the source,

    (iii) The initial and current quantity amount (e.g., milligrams, milliliters, grams, etc.),

    (iv) The toxin used and purpose of use, quantity, date(s) of the use and by whom,

    (v)   Where stored (e.g., building, room, and freezer),

    (vi) When moved from storage and by whom and when returned to storage and by whom including quantity amount,

    (vii) Records created under § 73.16 and 9 CFR part 121.16 (Transfers),

    (viii) For intra-entity transfers (sender and the recipient are covered by the same certificate of registration), the toxin, the quantity transferred, the date of transfer, the sender, and the recipient,

    (ix) Records created under § 73.19 and 9 CFR part 121.19 (Notification of theft, loss, or release), and

    (x)   If destroyed, the quantity of toxin destroyed, the date of such action, and by whom.

---

[a]Equivalent regulations for record-keeping for animal and plant select agents and toxins appear in § 331.17 of 7 CFR 331 and § 121.17 of 9 CFR 121.

Unlike nuclear materials, biological organisms have the ability to replicate (see Box 5-3).[7] Because a new culture can be prepared with as little as a single microorganism, an individual would need only a miniscule—and undetectable—amount from a single vial to establish a new culture and grow up large volumes of the agent in a matter of hours or a day. Therefore, determining that the number of vials is the same from one moment to another provides no guarantee that agents have not been removed from the laboratory since the original number of vials or tubes could remain the same while the agent itself has been removed. Also unlike nuclear materials, it is possible to completely inactivate BSAT materials: microorganisms can be autoclaved and toxins denatured so that they no longer pose a risk. As convenient as it might be to count vials, volumes, or number of organisms, it is not a biologically relevant means of inventory.

For these reasons, **the committee concluded that undue reliance on accounting practices, including counting vials, leads to false security and is counter-productive.**

So if counting vials or quantities is not a viable strategy for inventorying materials, what is? The committee strongly endorses a focus on accountability: specifically, *what* BSAT material is present, *when* it was obtained, *where* it is located, *who* has access, and *when*. It is also prudent that accountability not be limited to archived stocks but be extended to working materials as well, whether they are present in vials, Petri dishes, or laboratory animals. Accounting for the containers in which the cultures are maintained is neither feasible nor worthwhile. Accounting for access to select agents and toxins would be far more reliable and practical.

RECOMMENDATION 4: Because biological agents have an ability to replicate, accountability is best achieved by controlling access to archived stocks and working materials. Requirements for counting the number of vials or other such measures of the quantity of biological select agents (other than when an agent is transported from one laboratory site to another) should not be employed because they are both unreliable and counter-productive, yielding a false sense of security. A registered entity should record the identity of all biological select agents and toxins within that entity, where such agents are stored, who has access and when that access is available, and the intended use(s) of the materials.

It should be noted that Recommendation 4 makes a distinction between select agents—which have the capacity to replicate—and toxins—which do not. This recommendation, therefore, does not change the requirement to keep records on the amount of a toxin but does recommend that inventories for both

---

[7]Toxins do not generally present the same difficulty as they cannot, themselves, replicate.

BOX 5-3
Comparing Nuclear and Biological Materials

Nuclear Materials[a]

- Do not exist in nature in readily concentrated form appropriate for weapons
- Not living organisms
- Difficult and costly to produce
- Limited number of materials of concern
- Exist in forms of varying degree of concern
- Can often be detected at some distance using radiation detection equipment

Implications for Security—Nuclear Materials

- Since nuclear materials do not exist in concentrated form in nature, and are either the direct or indirect product of processes that are complex and expensive, security concerns surround both the materials as well as the processes used to produce them.
- Nuclear materials themselves range from nonlethal to highly lethal depending upon their specific properties at any point along the industrial process.
- Accounting for nuclear materials therefore occurs not only within each production process but also over the entire "nuclear fuel cycle" from initial stages of preparation (the "front end") through final stages of waste storage (the "back end").
- While most nuclear materials can be accounted for using quantitative measures, not all quantities and characteristics are known at all facilities holding nuclear materials. Further, a certain degree of material is "lost" through industrial processes and is termed "Materials Unaccounted For" (MUF).
- Some materials, particularly those at the back end, are highly lethal; anyone coming into unshielded contact with them would quickly receive a lethal dose of radiation. Such materials are therefore considered self-protecting. Security surrounding these materials is aided by these physical properties.
- Other, less lethal materials are heavily protected following a graded approach based on whether or not the materials are classified as weapons-grade and other physical properties that would affect ease of theft such as size, weight, quantity, and desirability for terrorist or other nefarious purposes.
- A wide variety of security measures have been developed and strengthened over several decades to ensure physical security of nuclear materials.
- A common security practice is the use of the two-person rule, implemented in different ways depending on the type of facility (research reactor, power reactor, nuclear processing facility, weapons facility).
- Personnel security measures also vary depending upon the facility and the types of activities being conducted at those facilities and generally follow a graded approach. These measures range from an internal background check conducted by the facility (research and test reactors) to FBI background checks as part of a Personnel Reliability Program (nuclear power plants) to full security clearances (defense facilities).
- Security measures are very costly and, in the case of nuclear materials in the civilian sector, regulated by an independent regulatory body, the Nuclear Regulatory Commission.

Biological Pathogens

- Generally found in the natural environment and often widely distributed globally
- Living organisms that replicate and can be inactivated or killed
- Easy and fairly cheap to produce
- Highly diverse
- Cannot be detected at a distance with available technologies

Implications for Security—Biological Pathogens

- Since biological pathogens do exist in nature, security concerns surround the facilities, processes, and pathogens that produce more purified and potentially dangerous strains and/or concentrations of pathogens.
- Pathogens range in their risk to human, animal, and plant health depending upon their concentrations, method and time of exposure, and other factors.
- Physical and personnel security of pathogens is built around protection of pathogens from intentional or accidental release into the environment.
- Accounting for organisms that replicate is a difficult challenge.
- Currently there is no graded approach to the security of pathogens on the list of select agents and toxins.
- A wide variety of security measures have been developed and strengthened over several decades to ensure security of stocks of pathogens.
- The two-person rule is used in some pathogen research settings when deemed appropriate.
- The SRA is the primary personnel security measure for those conducting research on select agents, but when this research is conducted in facilities where classified research is conducted, other security clearances may be employed.
- Security measures are very costly and regulated primarily by CDC and APHIS.

---

[a]The IAEA Safeguards Glossary, 2001 Edition, defines Fissionable Material as "in general, an isotope or a mixture of isotopes capable of nuclear fission. Some fissionable materials are capable of fission only by sufficiently fast neutrons (e.g., neutrons of a kinetic energy above 1 MeV). Isotopes that undergo fission by neutrons of all energies, including slow (thermal) neutrons, are usually referred to as fissile materials or fissile isotopes. For example, isotopes ^{233}U, ^{235}U, ^{239}Pu and ^{241}Pu are referred to as both fissionable and fissile, while ^{238}U and ^{240}Pu are fissionable but not fissile." Nuclear material is defined as "any source material or special fissionable material as defined in Article XX of [the Statute of the International Atomic Energy Agency, 1956]." <http://www-pub.iaea.org/MTCD/publications/PDF/nvs-3-cd/PDF/NVS3_prn.pdf>

select agents and toxins should include information about who has access to these materials, when, and for what intended purpose.

Accounting for BSAT Materials: The Personnel Aspect

The current security approach for select agent laboratories requires controlling who has access to BSAT materials. One of the most frequently suggested additional security measures to thwart an insider threat for BSAT laboratories is imposition of a "two-person" rule so that no one individual would have unsupervised access to BSAT materials. This rule was discussed, and often strongly criticized, in the public consultations sponsored by the NSABB and EO Working Group on Strengthening the Biosecurity of the United States. The directors of all the major BSL-4 laboratories in the United States argued that imposing a two-person rule could in fact decrease rather than improve safety in the laboratory and recommended using a video monitoring system instead (LeDuc et al. 2009).[8] The Defense Science Board found that neither video monitoring nor the two-person rule were effective as currently practiced, that the two-person rule had too many disadvantages to be considered as a standard practice, and that improvements in video monitoring should be made to increase its effectiveness (DSB 2009:20-22).

One of the problems with the debate over the two-person rule is that in practice there are a number of forms of such a rule, and it is not always clear that all participants in the debate mean the same thing by the term. The leading variants include:

- The model from the nuclear weapons "surety" program, in which one person performs the task and another person watches him or her

---

[8]The concern expressed by the BSL-4 laboratory directors in this regard was that:

> "Typically, the daily flow of work in a biocontainment laboratory involves ≥2 persons working near each other. Although the primary role of each person is to perform his or her work and not to monitor their coworker, there would still be ample opportunity to render aid if needed or to observe untoward activities. The risk issues of the 2-person rule arise when the normal activity in the laboratory is insufficient to satisfy the rule in every active area of the BSL-4 laboratory. These issues will occur frequently outside normal working hours, i.e., evenings, weekends, or holidays, and also during regular working shifts as project-specific tasks are completed asynchronously. The key implementation issue for a 2-person rule is when staff members are required to enter or extend time in containment solely to fulfill the security requirement. This personnel-centric approach has some serious shortcomings and in many circumstances may increase the safety risk for laboratory personnel. Effectively, the presence of an observer for the sole purpose of achieving a 2-person requirement would contribute to perceived time pressures, stress, distractions, and interruptions, all of which are factors identified by human performance management as error precursors" (LeDuc et al. 2009).

carry it out.[9] "One works, one watches" is the most costly option because of the need to have a person dedicated to observing rather than participating in other laboratory work, but it could be the one that in principle contributes the most to security because watching would be the only purpose of the second person. However, the visually subtle operations in biological research may make it difficult to detect troubling actions. This is the variant that was most strongly criticized and received the least support at the public consultations.

- The public consultations and the site visits and presentations to the committee suggested that the most common approach to the two-person rule currently being used is requiring that no one works in a laboratory alone. The second person is carrying out his or her own work but the presence of another individual is considered a deterrent and perhaps a way to detect illicit behavior as well. In almost all cases, the rule is applied for safety, not security reasons, since another individual is already present in the facility and able to detect and respond to emergency situations. Critics of this approach cite the uncertain effectiveness of simply having someone present when at least some acts could be quite readily concealed: the theft of a minute amount of BSAT material might elude even a careful observer.

- A variant described to the committee during its visit to the New England RCE at Harvard Medical School involves one person working and a second person assisting him or her. This procedure was implemented for safety reasons and is used during the more complicated aspects of working with BSAT material when another "pair of hands" is welcome. Since the procedure provides direct assistance to someone working in an admittedly inconvenient environment, it might be more acceptable to the research community, but it would still be an additional expense. Moreover, not all BSAT experiments require this level of support, so there are times when the presence of a second person is not justified.

- "No one works alone" is sometimes defined to mean that one person is in the laboratory and a second outside but nearby, with the ability to monitor the laboratory, with cameras and/or by regular, scheduled voice contact. If the individual on the outside detects a safety or

---

[9]"No lone individual shall have access to a nuclear weapon. During any operation that may require access to nuclear weapons, there is a minimum of two authorized persons, each capable of detecting incorrect or unauthorized procedures with respect to the task to be performed and familiar with applicable safety and security requirements. Two authorized personnel are physically positioned where they can detect incorrect or unauthorized procedures with respect to the task or operation being performed. When application of the two-person policy is required, it is enforced by the persons who constitute the team during the entire period they are accomplishing the task or operation assigned and until they leave the area within which the two-person policy is required" (Blaisdell 2001).

security problem or cannot contact the individual in the laboratory, he or she has the opportunity to solicit additional help or potentially enter the laboratory as warranted. This is separate from the question of having a dedicated monitoring program with video cameras.

- Although more often considered as part of physical security, the use of video recording or closed circuit television systems may also be used as part of personnel assurance programs to monitor activity in a BSAT laboratory or storage area. Some systems operate continuously and others rely on sensors and motion detectors to activate the cameras. Real-time monitoring, even if not continuous, is expensive and requires security personnel with sufficient training to recognize when someone in the laboratory is doing something illicit. Archived recordings have value for investigative purposes—or as a deterrent. According to one report, recordings are retained from a few weeks to a year, varying widely from laboratory to laboratory, which can limit their effectiveness as part of an investigation after an event has occurred or been alleged (DSB 2009:22).

The committee concluded that, when specifically indicated by a risk assessment, a rule that "no one works alone"—defined as one person conducting work in a laboratory while being in direct communication with a second person who can affect a rescue—should be in place. While there may be especially risky circumstances, such as certain procedures with nonhuman primates or work in especially challenging physical environments, most laboratory work has been designed to minimize safety risks. More important in the context of select agent research is that access to the laboratory facility is so restricted that any kind of accident might not be detected for some time. This is the motivation for workers in the laboratory being in regular contact with another individual. Since this is a safety measure, with only indirect security benefits, security is best maintained by regulating access—namely, requiring log-entry and -exit systems and electronic identification cards for all personnel.

SECURITY BASED ON RISK ASSESSMENT

Physical security is required of all facilities registered with the Select Agent Program.[10] Each facility must develop and implement a written security plan, which is reviewed by either CDC or APHIS as part of the initial and ongoing facility registration process. Because each facility is different in design,

---

[10]Department of Health and Human Services, 42 CFR 72 and 73; 42 CFR 1003 provide the final rule on the Possession, Use and Transfer of Select Agents and Toxins. The specific section addressing physical security is 42 CFR 73.11 (HHS 2005).

different physical security methods are required to address site-specific security requirements. Determination of which physical security measures to include in a site-specific plan is made based on "a site-specific risk assessment and must provided graded protection in accordance with the risk of the select agent or toxin, given its intended use" (HHS 2005:13306). In more detail, the security plan must:

1. Describe procedures for physical security, inventory control, and information systems control;
2. Contain provisions for the control of access to select agents and toxins;
3. Contain provisions for routine cleaning, maintenance, and repairs;
4. Establish procedures for removing unauthorized or suspicious persons;
5. Describe procedures for addressing loss or compromise of keys, passwords, combinations, etc. and protocols for changing access numbers or locks following staff changes;
6. Contain procedures for reporting unauthorized or suspicious persons or activities, loss or theft of select agents or toxins, release of select agents or toxins, or alteration of inventory records; and
7. Contain provisions for ensuring that all individuals with access approval from the HHS Secretary or APHIS Administrator understand and comply with the security procedures.

These regulations provide overall guidelines for the content of site-specific security plans; however, they are sufficiently broad to allow for variation in their implementation. While this variation has benefits, it also creates inconsistencies and confusion as facility operators, contractors, and inspectors attempt to determine whether specific security measures at individual facilities sufficiently adhere to these guidelines.

Variations in existing governing regulations create additional challenges for those designing and implementing site-specific, risk-based security plans. The long list of regulations applied to physical security includes the following:

- ARS 242.1-ARS Facility Design Standards (from the Agricultural Research Service)
- APHIS Security Design Standards
- *Biological Safety in Microbiological and Biomedical Laboratories* (CDC/NIH 2007)
- United Facilities Criteria (UFC) July 31, 2002
- UL 972 (12) Burglary Resistant Glazing Material-Standard Test Procedures-Smash and Grab Resistance

- UL 572 Standard Test Procedures for Bullet-Resistant Glazing
- ASTM 1233 (11) Burglary Resistant Glazing Material-Standard Test Procedures
- ASTM F 588 Resistance of Window Assemblies to Forced Entries
- ASTM F 476 Security of Swinging Door Assemblies
- ASTM F 842 Measurement of Forced Entry of Horizontal Sliding Door Assemblies
- ASTM F 1642 (21) Standard Test Method for Glazing and Glazing Systems Subject to Airblast Loadings
- NIH Draft Physical Security Design Guidelines for NIAID NBLs and RBLs, December 19, 2003
- Interagency Security Committee Security Design Criteria, for New Federal Office Buildings and Major Modernization Projects, May 28, 2001
- NIH Design Policy and Guidelines (spring 2003)
- Uniform Federal Accessibility Standards (UFAS) and/or the Americans with Disabilities Act (ADA)
- International Building Code, latest edition
- National Fire and Life Safety Codes, latest editions
- HHS (NIOSH) Publication No. 2002-139, Guidance for Protecting Building Environments from Airborne Chemical, Biological, or Radiological Attacks
- NIH Security Device Application Guideline #22230, latest edition
- *Laboratory Security and Emergency Response Guidance for Laboratories Working with Select Agents* (CDC 2002)
- U.S. Department of Energy, *A Manual for the Prediction of Blast and Fragment Loadings on Structures*
- 10 CFR 20.1801 and the Nuclear Regulatory Commission regulations on security of radioactive materials
- 42 CFR 73, including pathogens and toxins regulated by both HHS and USDA and non-overlap select agents of HHS
- Memorandum from the Secretary of the Department of Health and Human Services dated March 6, 2002, with 12 "Requirements for Securing Select Agents," Attachment 5
- AR-XX Military Police-Biological Agent Security Program
- UFC 4-010-01 DOD Anti-terrorism Standards

These varied regulations and guidelines lead to inconsistencies in application for a variety of reasons, reflective of the fact that facilities and regulations differ. On the facility side, Threat Risk Assessments, which are generally conducted by contractors, produce results that are unique to the facility and are influenced by the specific contractor. Therefore, the contractors recommend different solutions for similar situations. New and old facilities have different operational and physical infrastructures. Regulations themselves also vary from those that are prescriptive (e.g., doors must use a certain kind of lock) to those that have performance-based requirements (e.g., doors must be locked). They lack consistent fundamentals, and no detailed requirements for operational security needs are provided. This leaves facility owners and operators to decide which approaches are most appropriate, considering the likelihood that various inspectors might reach different conclusions or have different interpretations of the regulations.[11] Given these inconsistencies, facility designers and operators often err on the side of caution and implement physical security approaches that far exceed those warranted in a particular situation and are, therefore, inefficient. Addressing these inconsistencies and the problems they create would be beneficial from a security and cost-benefit perspective.

In order to address these inconsistencies, a minimum set of physical/technical standards should be established that interfaces with operational practices applicable to all relevant agencies. One means of doing this may be through a tiered approach to security based on mission. Further, a review process that verifies physical/technical security requirements would be helpful, particularly if combined with the establishment of an incident database to validate performance of established principles. Finally, updating the Threat Risk Assessments periodically (perhaps every 2-3 years and ideally in line with the inspection cycle to allow time for appropriate adjustments) would allow facility operators to adjust security measures accordingly, ensuring that security measures are continually implemented based on the most recent assessment of risk. This would allow for more cost-effective and consistent compliance with security needs and regulations.[12]

In conclusion, currently little guidance is provided by the Select Agent Program as to the definition and interpretation of minimum standards for physical security, leading to a significant difference in the design and associated cost of BSAT facilities. This often presents subsequent inspection and compliance challenges and does not necessarily ensure greater security.

RECOMMENDATION 6: The Select Agent Program should define minimum cross-agency physical security requirements, which recognize that facilities have unique risk-based security needs and associated design components, to assist facilities in meeting their regulatory obligations.

This recommendation is not intended to move away from the performance-based standards promulgated by the Select Agent Program but to provide additional guidance to the community on minimal standards. Moreover, any

---

[11]See below for a discussion of inspections and the need for inspector training to promote harmonization.

[12]Recommendation 9 suggests a mechanism for funding security upgrades.

stratification of the select agent list, as recommended above, may have implications for stratification of the physical security requirements as well. The Select Agent Program can further assist institutions in interpreting physical security requirements by establishing a hotline or other mechanism for rapid response in answering questions about interpretation of the standards.

EVALUATION

The committee believes that it is both appropriate and necessary to apply rigorous analytical methods to assess the mix of policies that promote both high-quality science and appropriate security. It is exceptionally difficult, however, to assess whether the policies are, in fact, achieving this optimal mix. Designing policies to prevent future terrorist activities and, later, understanding how and whether a program or programs prevented those activities and behaviors presents a particular evaluation challenge. After all, if the policies are successful, nothing bad will happen. Does this mean that the policies "worked," that is, that they prevented terrorists from gaining access to BSAT materials? Or does it mean that the threat of bioterrorism has been exaggerated, the policies were not needed, and funds were needlessly redirected from research? Or does it mean that we are worrying about the wrong aspects of threat?[13] Following from the difficulty in assessing the effectiveness of programs that will be successful if there is no obvious effect—other than the absence of another action—it is likewise difficult to assess whether the various costs associated with the program are appropriate.

Independent evaluation can provide useful information on how the Select Agent Program is implemented and identify important intended or unintended consequences of the program upon the research enterprise. Fortunately, there are several existing tools and techniques that could be applied in evaluating the program. The committee believes that any new policies intended to improve physical security and personnel reliability should be carefully evaluated, along with operation of the program overall. Relying on "dueling anecdotes" in not acceptable for establishing policy. We note that the National Research Council (NRC) report *Biotechnology Research in an Age of Terrorism* suggested a similar process for addressing another aspect of the efforts to reduce the risk of bioterrorism, that is the potential misuse of the knowledge, tools, and techniques that would result from BSAT (and other life sciences) research:

> The substantial expansion of funding for research in biodefense now in progress and anticipated suggests that it will be vital to assess how these new resources affect the conduct of research and to be ready to make timely

---

[13]For further discussion of these issues from different perspectives, see Danzig (2003), Altman et al. (2005), Leitenberg (2005), and Ostfield (2009).

adjustments. The monitoring should be done with the goal of suggesting ways to improve the [committee's proposed] system's operation and efficiency. But it should also include the possibility of proposing that parts of the system be overhauled or even eliminated if they prove ineffective or an impediment to important scientific research. (NRC 2004a:120)

The committee emphasizes that formal evaluation of the Select Agent Program is more than accumulation of metrics and demographic data. It is difficult to obtain quantitative measures of the impact of complex policies on difficult problems. Furthermore, only a portion of the information needed to gain a detailed understanding of the Select Agent Program will be quantifiable. While it is not within the scope of this committee to elaborate in detail or to design a specific program evaluation plan, several practices widely employed by governments at all levels, by the international development community, and by various types of private organizations provide a constructive framework applicable to the Select Agent Program. See, for example, discussion of the range of program evaluations methods in Wholey et al. (2004) and World Bank (2004).

Among key aspects of program evaluation, regardless of the specific evaluation methods selected, are identification of program goals and objectives, recognition of the need for both quantitative and qualitative data, and development of a specific evaluation design. Given that the Select Agent Program has evolved since its inception, data (when and where they exist) may be imperfect and not necessarily suited to addressing questions about the goals and objectives of the program. However, there are specific measures that can be taken to enhance the ability to understand the impact of the Select Agent Program on the promotion of high-quality science and maintenance of appropriate security.

As with many similar programs, the Select Agent Program has multiple goals and objectives. Therefore, it follows that evaluation of the program may be designed to understand its effectiveness and the impact of one or more of these goals and not others; or an aggregate analysis can be designed to understand the impact of the program as a whole. The first step in designing program evaluation is identifying which specific question or questions are to be addressed. Evaluation questions relevant to the BSAT research may include: What is the relative impact/effectiveness of various and/or combined physical security measures? What is the relative impact/effectiveness of the SRA process as a whole and in part? What are the costs and benefits of these measures and processes? What impact does the Select Agent Program have on the ability of laboratories to recruit new researchers to perform research on select agents and toxins and to retain expertise within the select agent community? How effective are the relevant management practices employed at BSAT laboratories? Answers to each of these questions may lead to different evaluation designs, but posing the questions to be addressed is a necessary first step.

Any program evaluation, regardless of the specific question(s) of interest, will require analysis of information, usually a combination of both quantitative and qualitative data. Since the inception of the Select Agent Program, the collection and retention of important data have increased, including gathering data on applicants for SRA clearance, SRA processing times, and the number of facilities registered as BSAT laboratories. Despite this progress, relatively little data have been collected on the Select Agent Program as a whole. It is still difficult to know with any degree of certainty how much the different aspects of the program cost or how many researchers may or may not choose to work with select agents and toxins because of concerns about the SRA process. Going forward, more and varied types of data will be required to understand—either in part or in full—the impact of the Select Agent Program on the research community, on the type and quality of research undertaken, and ultimately on the safety and security of those working with select agents and toxins and of the surrounding community. A baseline survey may be an effective means of gathering initial data that would establish a more clear understanding of key aspects of the Select Agent Program and inform future evaluation efforts.

There are numerous means by which program evaluation experts can and do collect relevant data. Specific decisions regarding data collection are made based on cost, available personnel resources, ease and/or difficulty of data collection, and specific demands of the chosen program evaluation design. However, certain methodologies are common to many designs, such as surveys, interviews, and the addition of new questions to existing forms and/or reporting mechanisms. Most program evaluations employ a combination of these techniques to acquire the relevant and necessary data. In the BSAT context, program evaluation will require all segments of the research and regulatory community to participate in data collection, as there is no central location that has access to all of the relevant perspectives. Universities are in a position to collect data on the research and career decisions of graduate students, the regulatory and implementing agencies are in a position to collect data on certain costs and processing times, private biosecurity companies and facilities are in a position to collect data on physical security costs, and program evaluation experts are in a position to bring their experience with evaluating complex systems to bear on the task of collecting qualitative data. Moving beyond anecdotal information to data collection and analysis will require coordinated effort and participation.

Some of the most important studies to undertake may also be among the most difficult. For example, what can we learn about the research that is *not* conducted as a result of some aspect of the Select Agent Program? What are the missed opportunities and discoveries that are delayed or not achieved because of disincentives to select agent research? How do the financial, personnel, and security limitations impact which experiments are conducted and by whom? While efforts to evaluate the BSAT program, particularly in the initial stages, will present challenges to evaluation experts and many in the BSAT research community, there is a broad and growing body of expertise upon which to draw. Significant progress has been made in recent years and decades to improve evaluation and the results it can offer to stakeholders. A broad array of evaluation designs and techniques now exists to provide decisionmakers with the information and analyses necessary to make appropriate policy and legislative decisions that can improve the effectiveness of programs. Given the evolution of the Select Agent Program, and the genuine concerns of lawmakers, scientists, and citizens alike, it would be beneficial to undertake a series of evaluation studies to address those questions deemed relevant and appropriate.

> **RECOMMENDATION 7: Independent evaluation of the Select Agent Program should be undertaken to assess the relative benefits for achieving security, to consider the consequences of the program (intended and unintended) on the research enterprise, and to provide useful data about the Select Agent Program. Such evaluation, which may be coordinated through the BSAT Advisory Committee, should be provided with dedicated funding.**

The committee believes that it would be effective for evaluation to be coordinated through the BSATAC, since it will provide an important advisory role across the entire Select Agent Program, including the connections between regulatory agencies, funding agencies, and entities performing BSAT research. In addition, if BSATAC is housed within OSTP, the results of evaluation would be directly available to those within and beyond the Select Agent Program who are able most effectively to restructure programs and implement interagency coordination.

TRAINING OF INSPECTORS

All select agent laboratories undergo regular inspections by CDC or APHIS, whether academic, commercial, or government and whether for research or public health. Routine inspections occur every three years with additional inspections undertaken when there is a significant change in an entity's registration or in response to concerns (see Chapter 2 for more information). These inspections involve both extensive review of records and a multi-day visit to the facility during which inspectors verify the accuracy of records, inspect the security and safety components of the facility, and interview personnel.

In addition to the inspections by agencies with statutory responsibility for the Select Agent Program, many funding agencies—including DHS and the Department of Defense—conduct their own inspections on research and facilities they support. Other federal agencies also have responsibility for overseeing aspects of the facility and may conduct inspections. Finally, some state and local authorities inspect facilities within their jurisdiction.

Close coordination between CDC and APHIS in the Select Agent Program has served the research community well and should be expanded to include other government agencies with an involvement in BSAT research. Specifically, the committee encourages coordination and consolidation so that entities with select agent research sponsored and/or regulated by different federal agencies are not subject to very different and possibly conflicting guidance and regulations or to duplicative inspections. The committee is hopeful that the conversation begun through the interagency EO Working Group may have tangible benefits in the coordination of regulations and practices among the various U.S. government agencies that have some responsibility for BSAT research.

The current statutory authority of agencies with responsibility for the Select Agent Program will likely make it difficult to achieve a perfectly streamlined system of oversight and inspection. For example, the committee recognizes that it will be difficult for one agency to defer completely to the inspection conducted by a different agency because of Congressionally mandated responsibilities and different areas of focus. However, there are steps that can be taken to streamline different inspections, such as by having them coordinated in time and operating from a common set of fundamental principles and policies. In addition, it is critical to ensure that the requirements of multiple agencies are not contradictory. At the present time, the benefit of inspections is compromised by confusion caused by overlapping and sometimes conflicting guidance, as well as inconclusive findings from a variety of agencies and bodies.

The multiplicity of oversight and regulatory functions leads to significant expenditure of time and resources by the select agent entity in its preparing for, participating in, and responding to the inspection. The intent is that inspections provide an opportunity to consider important aspects of security and compliance issues. In fact, the committee heard from some members of the community that inspections conducted by well-trained and experienced inspectors were quite helpful and educational.

Nevertheless, complaints about the nature of some inspections have arisen (e.g., Box 5-4). Members of the community have cited the increasingly bureaucratic nature of some inspections, with expanding focus on the technical letter of the regulation without regard to the spirit of the regulation and its intended objective. For example, one entity cited an example of an inspector who was unwilling to accept an "Emergency Response Plan" when an "Incident Response Plan" was listed in the inspection checklist; simply changing the title on the document satisfied the inspector.

Concerns have been raised that inspectors have not had the technical knowledge needed to understand the specific nature of various risks and have been reluctant to seek guidance from knowledgeable colleagues at CDC or APHIS. While the Select Agent Program is structured to provide several levels of support when necessary, it appears that not all inspectors make use of these opportunities, resulting in inspection reports that are not fully informed by the scientific issues impacting security and safety of the laboratory.

BOX 5-4
The Challenge of Compliance

The Southwest Foundation for Biomedical Research has been working with select agents since 1996 and housed the only operational academic BSL-4 laboratory from 2000 until the opening of the laboratory at the University of Texas Medical Branch at Galveston in 2004.

The laboratory's research on vaccines, therapeutics, and pathogen detection spans multiple sponsors and agencies. This means that their facilities have been subject to inspection and oversight by an alphabet of government agencies. In recent years, the laboratory has been visited by representatives from CDC, HHS, USDA, DHS, Government Accountability Office (GAO), Environmental Protection Agency (EPA), Department of State, Department of Commerce, and the Department of Transportation, among others.

Multiple levels of oversight can cause conflict and confusion. For example, a 2008 GAO report cited the presence of a window in one laboratory as a security risk (GAO 2008), even though the placement of the window had been recommended by other maximum-containment labs and the windows were high above the ground, constructed with bulletproof glass, and equipped with bars (GAO 2009a). What is an entity to do when multiple authorities are not working from the same rulebook?

According to Dr. Joan L. Patterson, who chairs the Department of Virology and Immunology, the foundation estimates that the laboratory spends nearly one-quarter of its time and resources related to inspections.

The foundation has recently decided to discontinue its relationship with DHS, because of a combination of poorly trained inspectors and required compliance that far exceeded the funds provided. The foundation found that inspectors were requiring compliance with procedures that were not in line with recommended biosafety practices, were asking questions the foundation deemed inappropriate, and were asking for access to the BSL-4 laboratory, even though the inspectors had no familiarity or training in that environment. These aspects of inspection were deemed outside the norm of what is reasonable and prudent—and led to the conclusion that continued DHS funding did not outweigh the costs and challenges of complying with contradictory rules and expectations.

Much of the concern may stem from the fact that some inspectors are not sufficiently familiar with the nature of BSAT research. Although the Select Agent Program seeks to hire inspectors with scientific experience, including work in select agent laboratories, there are others hired who come from a biosafety or regulatory background but without an understanding of the select agent laboratory environment. These challenges are even more severe for those government agencies that do not focus on select agents, such as the community of state and local health officials who have rarely encountered a select agent facility but may have a responsibility for inspecting them.

Because of the critical role that inspections play in monitoring the safety and security of select agent facilities and identifying areas for improvement, it is necessary that those individuals who conduct the inspections have the proper background, training and support. Currently, not all inspectors have this preparation. It is also important that inspectors from the many agencies that oversee select agent research or facilities receive similar training to facilitate harmonization in the application of select agent regulations and inspection guidelines.

RECOMMENDATION 8: Inspectors of select agent laboratories should have scientific and laboratory knowledge and experience, as well as appropriate training in conducting inspections specific to BSAT research. Inspector training and practice should be harmonized across federal, state, local, and other agencies.

Moreover, appropriately trained and experienced inspectors should be supported by their agencies in exercising informed discretion. Inspectors should be encouraged to utilize their experience and training to apply consistent regulations to variable environments and to solicit input from others where appropriate.

Although the committee does not make a specific recommendation in this regard, there may be a role for a certification process that will identify those inspectors who have the appropriate training and experience for inspecting select agent laboratories.

FUNDING SECURITY AND COMPLIANCE COSTS

Security and compliance procedures called for under the Select Agent Program can be significant, with costs substantially higher than for similar laboratory facilities. Security guards, cameras, access card readers, biometric identification technologies, alarms, lockable freezers and incubators, and other security measures all add to the cost of operating a select agent laboratory. Specialized equipment installed and maintained for biosafety purposes, such as air filtration systems, biosafety cabinets, decontamination showers, changing rooms, and the like have high initial and maintenance costs.

Construction of secure laboratories where select agent research will be conducted is often funded by grants specific for that purpose. For example, the National Institute of Allergy and Infectious Diseases (NIAID) is currently in the process of building a series of national and regional biocontainment laboratories (NBLs and RBLs). With tens to hundreds of millions of dollars invested in each of the two NBLs and 13 RBLs operational or under construction, as well as a network of 10 RCEs for Biodefense and Emerging Infectious Diseases, NIAID has made a substantial contribution to the nation's biosecurity infrastructure.

But it is not only construction costs that make select agent facilities expensive. They have significant ongoing security and safety sustainment costs that far exceed the indirect costs that grantee institutions receive to cover the cost of facilities, maintenance, and operations. Although the committee did not have an opportunity to collect detailed information on security and compliance costs (see Box 5-5 for an example), comments on implementation of the select agent rules referenced in the *Federal Register* provide some indication. For example, annual operations and maintenance costs that were cited range from $100,000 to $700,000, with startup costs in the range of $1-4 million (HHS 2005). One university reported $300,000 in security improvements for electronic card access, alarm systems, and security cameras—not including additional recordkeeping and personnel requirements. Another institution cited a figure of $400,000 for a single BSL-3 laboratory. This is in line with reported cost of $130,000 to increase security at Louisiana State University and $130,000 to inventory and secure pathogens at Northern Arizona University (Wilkie 2004). The final Regulatory Impact Analysis to inform the select agent regulations, released in 2005, cited an annualized cost of the select agent regulations of $16 million, with annualized costs per facility of $15,300-$170,000 (HHS 2005). USDA also evaluated the cost of compliance. At the Ames, Iowa, Na-

---

BOX 5-5
The Challenge of Funding the Operation
of a Select Agent Laboratory

One of the committee's site visits was to the National Center for Biodefense and Infectious Diseases at the Prince William Campus of George Mason University (GMU), which is currently constructing a Biomedical Research Laboratory (BRL) that will conduct research with select agents and other materials at the BSL-2 and BSL-3 safety levels.

GMU security officials mentioned an intent to dedicate 13 security officers to provide round-the-clock protection to this single 52,000 square foot laboratory building with a staff of fewer than 70 people. In comparison, the entire rest of the Prince William Campus has five dedicated security officers, even though the 124-acre campus serves more than 4,000 students in its classrooms, laboratories, libraries, recreation, and other facilities.

Overall, GMU estimates that the operation of the BRL will be 2½-3 times the cost of other laboratory facilities. In addition to security guards, additional costs include maintaining air handling systems, maintenance within the BSL-3 environment, and ongoing costs for cameras, security systems, and other activities unique to the BRL—which will be the most secure building at the university.

tional Veterinary Services Laboratory operated by USDA, for example, the cost of security upgrades implemented in 2002 was over $550,000 (USDA 2005). Although the needs and expenses will depend upon the current status of the facility and the site-specific risk assessment—which itself can cost $20,000 or more to conduct—the expenses can be a significant burden.

In addition to the financial cost of security requirements, personnel affiliated with select agent research facilities spend significant amounts of time ensuring compliance. This includes both additional time dedicated to screening, recordkeeping, and reporting as well as time spent preparing for, participating in, and responding to the required inspections. This time will be even greater for facilities that conduct research overseen by more than one agency. A Stimson Center survey found that researchers reported spending approximately four hours per week on select agent compliance issues, more than the time spent on regulatory demands for other issues (Fischer 2006).

Regardless of the specific resources dedicated to compliance and reporting, there is little dispute that the amount is significant: tens of thousands of dollars per year at even the smallest facilities. To be sure, the costs of security compliance pale in comparison to the astronomical costs of an incident and the secondary effects on research if security were to be compromised (Ekboir 1999; HHS 2005), but these operational and maintenance costs are real and need to be funded. How will these additional security and compliance bills be paid on an ongoing basis? If the institution will be relying, in part, on indirect costs from research conducted at the facility, what will happen if those research grants are not available?

Some select agent facilities provide all of the necessary support for research, personnel, and materials through infrastructure grants. For example, the New England RCE at Harvard Medical School, which the committee visited, is operating almost entirely on the RCE grant from NIAID. The center does not charge investigators for work conducted in its BSL-3 lab, including costs for the facility and its staff. Other laboratories operate as a fee for service in that investigators' grants and other parties are charged for the direct costs of the research including laboratory staffing, perhaps even including a select agent surcharge. Other facilities may be limited to internally supported research with all of the security and compliance costs dependent upon indirect cost recovery and additional institutional support.

While the challenges of sustainable funding for scientific research go far beyond select agent research and this report, the implications are more troubling in the case of select agent research. It is not acceptable, either for the institution or for safety and security, to diminish appropriate and necessary risk-based security procedures and resources, regardless of the availability of funding for the facility. Host institutions, having to provide the difference, may choose to reduce their cost by understaffing the facility, hiring external contractors where a third party takes responsibility for key functions, or diverting funds planned to support scientific research to pay for security responsibilities. These are not sustainable solutions and raise risks.

The committee, therefore, urges federal agencies that fund BSAT research to establish dedicated funding for ongoing security and compliance responsibilities associated with this type of research. This is an essential obligation, and no facility should operate without appropriate security measures in place. Funds to support security and compliance should be from a separate source to avoid diminishing the already limited support for research and should be available on a continuing and competitive basis for the life of the facility.

RECOMMENDATION 9: Because of considerable security and compliance costs associated with research on biological select agents and toxins, federal agencies funding BSAT research should establish a separate category of funding to ensure sustained support for facilities where such research is conducted.

It is expected that these costs will be site specific and subject to change as security standards, risks, and successful practices evolve. The funding arrangements should also include a mechanism for supporting facility upgrades and implementing evolving standards and practices. In all cases, it will be important that these costs not be short-changed. In addition, the specific mechanism for providing such support will depend upon the nature of the laboratory and the funding source; for example, it may make sense to provide one mechanism for supporting the continued operation of federally funded facilities such as NBLs, RBLs, and RCEs and a different mechanism for investigator-initiated research grants.

Although this type of funding structure may be unusual for biomedical research laboratories, it is not uncommon for funding those areas of science where central infrastructure plays an important role. Primate research centers, telescopes, and the academic research fleet all have funding models in which operating costs are broken out as a separate direct expense, often from a separate account so that operations do not compete directly with science funding. The U.S. Academic Research Fleet, for example, divides the total operating expenses by the number of days the ship is at sea and charges this rate through ship operating proposals submitting to granting agencies (see Box 5-6). In this way, the granting agency pays for operating expenses directly and subject to the needs of the research projects but without relying upon research grants.

BOX 5-6
Funding Model for the Academic Research Fleet

The U.S. Academic Research Fleet provides an interesting model by which a federal agency takes responsibility for not only helping support the initial construction of a key element of research infrastructure, but also the continued maintenance of that research asset.

More than half of the 22 research vessels that are part of the fleet are owned by the Navy or the National Science Foundation. These agencies support initial construction of the ships as well as operating costs, even though the ships are operated by extramural academic institutions under a Charter Party or Cooperative Agreement. The operating entity is responsible for providing a crew for the ship and providing support on shore. The ships become a part of the University-National Oceanographic Laboratory System (UNOLS), a consortium of 61 academic institutions that serves as an advisory committee to the federal agencies.

UNOLS works with relevant federal agencies to develop the most efficient and cost-effective schedule of science cruises for each ship. Once the schedule is determined, the ship operator divides the total annual operational costs for the ship by the number of days at sea to calculate a day rate. When researchers apply for funding to conduct research on the ship, the day rate associated with that research project becomes the subject of a separate funding request from the ship operator to the agency supporting the research activities.

So, for example, if a particular research project will require $1 million in research support and 30 days of time on a ship with a day rate of $25,000 per day, that will result in one proposal from the researcher to support the $1 million science budget and a separate request from the ship operator for $750,000 to fund ship operations during the 30-day cruise.

References

AAAS (American Association for the Advancement of Science). 2008. *Professional and Graduate-Level Programs on Dual Use Research and Biosecurity for Scientists Working in the Biological Sciences: Workshop Report.* Washington, DC: AAAS. Available at <http://cstsp.aaas.org/files/AAAS_workshop_report_education_of_dual_use_life_science_research.pdf>.

AAAS. 2009. *Biological Safety Training Programs as a Component of Personnel Reliability: Workshop Report.* Washington, DC: AAAS. Available at <http://cstsp.aaas.org/files/AAAS%20Biosafety%20report.pdf>.

ABSA (American Biological Safety Association). 2009. ABSA Comments to the Working Group on Strengthening the Biosecurity of the United States Following the Public Meeting May 13-14, 2009. Available at <http://www.absa.org/pdf/090530ABSAcommentsBWG.pdf>.

Altman, Sidney, Bonnie L. Bassler, Jon Beckwith, Marlene Belfort, Howard C. Berg, Barry Bloom, Jean E. Brenchley, Allan Campbell, R. John Collier, Nancy Connell, Nicholas R. Cozzarelli, Nancy L. Craig, Seth Darst, Richard H. Ebright, Stephen J. Elledge, Stanley Falkow, Jorge E. Galan, Max Gottesman, Richard Gourse, Nigel D.F. Grindley, Carol A. Gross, Alan Grossman, Ann Hochschild, Martha Howe, Jerard Hurwitz, Ralph R. Isberg, Samuel Kaplan, Arthur Kornberg, Sydney G. Kustu, Robert C. Landick, Arthur Landy, Stuart B. Levy, Richard Losick, Sharon R. Long, Stanley R. Maloy, John J. Mekalanos, Frederick C. Neidhardt, Norman R. Pace, Mark Ptashne, Jeffrey W. Roberts, John R. Roth, Lucia B. Rothman-Denes, Abigail Salyers, Moselio Schaechter, Lucy Shapiro, Thomas J. Silhavy, Melvin I. Simon, Graham Walker, Charles Yanofsky, and Norton Zinder. 2005. An open letter to Elias Zerhouni. *Science* 307(5714, March 4):1409-1410.

Atlas, Ronald M. 1999. "Statement of American Society for Microbiology Task Force on Biological Weapons Control Co-Chair Dr. Ronald M. Atlas." Testimony before the Commerce Committee's Subcommittee on Oversight & Investigations' Hearing on Assessing the Adequacy of Federal Law Relating to Dangerous Biological Agents. May.

Auchincloss, Hugh. 2007. "Protecting the Public Health: The Importance of NIH Biodefense Research Infrastructure." Testimony before the U.S. House of Representatives Committee on Energy and Commerce, Subcommittee on Oversight and Investigations. October 4.

Berry, Christopher M., Paul R. Sackett, and Shelly Wiemann. 2007. A review of recent developments in integrity test research. *Personnel Psychology* 60:271-301.

Besser, Richard E. 2007. "Oversight of Select Agents by the Centers for Disease Control and Prevention." Testimony before the subcommittee on Oversight and Investigations, Committee on Energy and Commerce, U.S. House of Representatives. October 4.

Blaisdell, MG Franklin J. 2001. "Nuclear Weapons Security." Statement to the U.S. Senate Arms Service Strategic Subcommittee. December 13.

Borum, Randy. 2004. *Psychology of Terrorism.* Tampa: University of South Florida.

Brant, David, and Michael G. Gelles. 2009. Discussion Paper on Insider Threat Assessments: Insight into Developing a Secure Workforce, Identifying Current Risks, and Developing Mitigation Strategies (draft). Deloitte Consulting.

Burnett, LouAnn. 2009. Presentation to the Committee. June 29.

Burrelli, David F. 2009. *"Don't Ask, Don't Tell:" The Law and Military Policy on Same-Sex Behavior.* Washington, DC: Congressional Research Service.

Butcher, James N., Deniz S. Ones, and Michael Cullen. 2006. "Personnel screening with the MMPI-2." In *MMPI-2: A Practitioner's Guide* (James N. Butcher, ed.), pp. 381-406. Washington, DC: American Psychological Association.

Carus, W. Seth. 2001. *Bioterrorism and Biocrimes: The Illicit Use of Biological Agents Since 1900* (February 2001 revision). Working Paper from the Center for Nonproliferation Research. Washington, DC: National Defense University.

Cascio, Wayne F. 2009. *Managing Human Resources: Productivity, Quality of Work Life, Profits*, 8th ed. New York: McGraw-Hill.

CDC (Centers for Disease Control and Prevention). 2002. Laboratory security and emergency response guidance for laboratories working with select agents. *Morbidity and Mortality Weekly Report* 51(No. RR-19, December 6):1-6.

CDC. 2009. Facts about the Laboratory Response Network. Available at <http://www.bt.cdc.gov/lrn/factsheet.asp>.

CDC/APHIS (CDC/Animal and Plant Health Inspection Service). 2007. *Select Agents and Toxins: Security Information Document.* Atlanta, GA, and Riverdale, MD: CDC and APHIS. Available at <http://www.selectagents.gov/resources/Security%20Information%20Document.pdf>.

CDC/NIH (CDC/National Institutes of Health). 2007. *Biosafety in Microbiological and Biomedical Laboratories*, 5th ed. (L. Casey Chosewood and Deborah E. Wilson, eds.). Washington, DC: U.S. Government Printing Office.

CEN (European Committee for Standardization). 2008. *International Laboratory Biorisk Management Standard.* CWA 15793. Brussels: CEN.

Concept Systems Inc. 2008. *RCE Program Interim Evaluation: Report on the First Five Years of the Regional Centers of Excellence for Biodefense and Emerging Infectious Diseases Research (RCE) Program.* Bethesda, MD: National Institute of Allergy and Infectious Diseases, Division of Microbial and Infectious Diseases.

Crenshaw, Martha. 1981. The causes of terrorism. *Comparative Politics* 13(4):379-399.

Crowley, M. Colleen. 2009. Presentation to the Committee. August 11.

Cullen, Michael J., Deniz S. Ones, Chockalingam Viswesvaran, Shelly Drees, and Kathryn Langkamp. 2003. A Meta-Analysis of the MMPI and Police Officer Performance. In *Personality and Work Behaviors of Police Officers* (S. Spillberg and D.S. Ones, co-chairs). Symposium conducted at the annual meeting of Society for Industrial and Organizational Psychology, Orlando, FL.

Damos, Diane L. 2009. Presentation to the Committee. August 10.

Danzig, Richard. 2003. *Catastrophic Bioterrorism—What Is to Be Done?* Washington, DC: Center for Technology and National Security Policy, National Defense University.

Demmin, Gretchen L. 2007. Biosurety. In *Textbook of Military Medicine: Medical Aspects of Biological Warfare* (Martha K. Lenhart, ed.). Washington, DC: Borden Institute, Walter Reed Army Medical Center.

Department of the Army. 2008. Nuclear and Chemical Weapons and Materiel: Biological Surety. Regulation 50–1. Washington, DC: Department of the Army.

Department of State. 2009a. UN Security Council Resolution 1540. Available at <http://www.state.gov/t/isn/c18943.htm>.

Department of State. 2009b. State Sponsors of Terrorism. Available at <http://www.state.gov/s/ct/c14151.htm>.

DHB (Defense Health Board). 2009. *Defense Health Board Task Force Review of the Department of Defense Biodefense Infrastructure and Biological Research Portfolio*. Memorandum for the Surgeon General of the Army. April 29. Available at <http://www.health.mil/dhb/recommendations/2009/2009-01.pdf>.

DOJ (Department of Justice). 2006. *The Attorney General's Report on Criminal History Background Checks*. Available at <http://www.usdoj.gov/olp/ag_bgchecks_report.pdf>.

DSB (Defense Science Board). 2009. *Department of Defense Biological Safety and Surety Program*. Available at <http://www.acq.osd.mil/dsb/reports/2009-05-Bio_Safety.pdf>.

Ekboir, Javier M. 1999. *Potential Impact of Foot-and-Mouth Disease in California: The Role and Contribution of Animal Health Surveillance and Monitoring Services*, Davis, CA: Agricultural Issues Center, Division of Agriculture and Natural Resources, University of California, Davis.

FASEB (Federation of American Societies for Experimental Biology). 2009. FASEB Statement on Dual Use Research and Biosecurity Education. March 5. Available at <http://opa.faseb.org/pdf/2009/FASEB_Statement_on_Dual_Use_Education.pdf>.

FASEB and AAMC (Association of American Medical Colleges). 2009. Letter to U.S. Working Group on Strengthening the Biosecurity of the United States. May 29, 2006. Available at <http://opa.faseb.org/pdf/2009/Biosecurity_FASEB_AAMC_5.29.09.pdf> and <http://www.aamc.org/advocacy/library/research/corres/2009/052909.pdf>.

FBI (Federal Bureau of Investigation). 2008. Science Briefing on the Anthrax Investigation: Opening Statement by Dr. Vahid Majidi. Available at <http://www.fbi.gov/page2/august08/anthraxscience_081808.html>.

Fein, Robert A., Bryan Vossekuil, and Gwen A. Holden. 1995. *Threat Assessment: An Approach to Prevent Targeted Violence*. Washington, DC: National Institute of Justice.

Fein, Robert A., and Bryan Vossekuil. 2009. Presentation to the Committee. June 29.

Fischer, Julie E. 2006. *Stewardship or Censorship? Balancing Biosecurity, the Public's Health, and the Benefits of Scientific Openness*. Washington, DC: The Henry L. Stimson Center.

GAO (Government Accountability Office). 2007. *High Containment Biosafety Laboratories: Preliminary Observations on the Oversight of the Proliferation of BSL-3 and BSL-4 Laboratories in the United States*. GAO-08-108T. Testimony before the Subcommittee on Oversight and Investigations, Committee on Energy and Commerce, U.S. House of Representatives.

GAO. 2008. *Biosafety Laboratories: Perimeter Security Assessment of the Nation's Five BSL-4 Laboratories*. GAO-08-1092. Washington, DC: GAO.

GAO. 2009a. *Biosafety Laboratories: BSL-4 Laboratories Improved Perimeter Security Despite Limited Action by CDC*. GAO-09-851. Washington, DC: GAO.

GAO. 2009b. *Personnel Security Clearances: An Outcome-Focused Strategy Is Needed to Guide Implementation of the Reformed Clearance Process*. GAO-09-488. Washington, DC: GAO.

GAO. 2009c. *High-Containment Laboratories: National Strategy for Oversight Is Needed*. GAO-09-574. Washington, DC: GAO.

George, Alexander L., and Andrew Bennett. 2005. *Case Studies and Theory Development in the Social Sciences*. Cambridge, MA: The MIT Press.

Gottron, Frank, and Dana Shea. 2009. *Oversight of High-Containment Biological Laboratories: Issues for Congress*. Washington, DC: Congressional Research Service.

Gronvall, Gigi Kwik. 2008. Improving the Select Agent Program. *Bulletin of the Atomic Scientists*. Web Edition (October 29). Available at <http://www.thebulletin.org/web-edition/features/improving-the-select agent-program>.

Hafer, Nathaniel, Cheryl J. Vos, Karen McAllister, Gretchen Lorenzi, Christopher Moore, Kavita M. Berger, and Michael Stebbins. 2008. How scientists view law enforcement. New survey of researchers tells us how to help the communities communicate. *Science Progress* (December 22). Available at <http://www.scienceprogress.org/2008/12/science-and-law-enforcement/>.

Herbig, Katherine L. 2008. *Changes in Espionage by Americans: 1947-2007.* Technical Report 08-5. Monterey, CA: Defense Personnel Security Research Center.

Herbig, Katherine L., and Martin F. Wiskoff. 2002. *Espionage Against the United States by American Citizens 1947–2001.* Technical Report 02-5. Monterey, CA: Defense Personnel Security Research Center.

HHS (Department of Health and Human Services). 2005. "42 CFR 72 and 73 and 42 CFR Part 1003: Possession, Use, and Transfer of Select Agents and Toxins; Final Rule" (FR Doc. 05-5216). *Federal Register* 70(52, March 18), pp. 12294-13325.

HHS. 2009a. "42 CFR Part 73: Possession, Use, and Transfer of Select Agents and Toxins; Proposed Addition of SARS-Associated Coronavirus (SARS-CoV)." *Federal Register* 74(132, July 13), pp. 33401-33403.

HHS. 2009b. "42 CFR Part 73: Possession, Use, and Transfer of Select Agents and Toxins—Chapare virus." (FR Doc. E9-19901). *Federal Register* 74(159, August 19), pp. 41829-41831.

IOM (Institute of Medicine). 2000. *To Err Is Human: Building a Safer Health System.* Washington, DC: National Academy Press.

IOM. 2009. *Live Variola Virus: Considerations for Continuing Research.* Washington, DC: National Academies Press.

IOM/NRC (National Research Council). 2009. *Effectiveness of National Biosurveillance Systems: BioWatch and the Public Health System: Interim Report.* Washington, DC: National Academies Press.

Kelly, Terrence K., Peter Chalk, James Bonomo, Brian A. Jackson, and Gary Cecchine. 2004. *The Office of Science and Technology Policy Blue Ribbon Panel on the Threat of Biological Terrorism Directed Against Livestock: Conference Proceedings, Washington, DC, December 8-9, 2003.* Arlington, VA: RAND Science and Technology.

Kiehlbauch, Julia A., Robbin S. Weyant, Freeda Isaac, and Michael J. Firko. 2009. "The Federal Select Agent Program." Presentation to the National Academies Committee on Scientific Milestones for the Development of a Gene-Sequence-Based Classification System for Oversight of Select Agents. September 3.

LeDuc, James W., Kevin Anderson, Marshall E. Bloom, Ricardo Carrion, Jr., Heinz Feldmann, J. Patrick Fitch, Joan B. Geisbert, Thomas W. Geisbert, Michael R. Holbrook, Peter B. Jahrling, Thomas G. Ksiazek, Jean Patterson, and Pierre E. Rollin. 2009. Potential impact of a 2-person security rule on biosafety level 4 laboratory workers. *Emerging Infectious Diseases Online Report* 15(7, July). Available at <http://www.cdc.gov/EID/content/15/7/e1.htm>.

Leitenberg, M. 2005. *Assessing the Biological Weapons and Bioterrorism Threat.* Carlisle Barracks, PA: Strategic Studies Institute, U.S. Army War College.

Macrina, Francis L. 2005. *Scientific Integrity*, 3rd ed. Washington, DC: ASM Press.

Morgan, Gareth. 1997. *Images of Organization.* Thousand Oaks, CA: Sage Publications.

Murphy, Kevin. 2009. "Assessment Challenges." Presentation to the Committee. June 29.

NAE (National Academy of Engineering). 2009. *Ethics Education and Scientific and Engineering Research: What's Been Learned? What Should Be Done? Summary of a Workshop.* Washington, DC: National Academies Press.

NELP (National Employment Law Project). 2009. *A Scorecard on the Post-9/11 Port Worker Background Checks: Model Worker Protections Provide a Lifeline for People of Color, While Major TSA Delays Leave Thousands Jobless During the Recession.* Available at <http://nelp.3cdn.net/2d5508b4cec6e13da6_upm6b20e5.pdf>.

NIAID (National Institute of Allergy and Infectious Diseases). 2007. *NIAID Strategic Plan for Biodefense Research, 2007 Update.* Bethesda, MD: National Institutes of Health.

NIAID. 2008. *NIAID: Planning for the 21st Century, 2008 Update.* NIH Publication No. 08-4753. Bethesda, MD: National Institutes of Health.

NRC (National Research Council). 2003. *The Polygraph and Lie Detection.* Washington, DC: National Academies Press.

NRC. 2004a. *Biotechnology Research in an Age of Terrorism.* Washington, DC: National Academies Press.

NRC. 2004b. *Seeking Security: Pathogens, Open Access, and Genomic Databases.* Washington, DC: National Academies Press.

NRC. 2005. *Bridges to Independence: Fostering the Independence of New Investigators in Biomedical Research.* Washington, DC: National Academies Press.

NRC. 2006. *Globalization, Biotechnology, and the Future of the Life Sciences.* Washington, DC: National Academies Press.

NRC. 2007a. *Technical Input on the National Institutes of Health's Draft Supplementary Risk Assessments and Site Suitability Analyses for the National Emerging Infectious Diseases Laboratory, Boston University: A Letter Report.* Washington, DC: National Academies Press.

NRC. 2007b. *Science and Security in a Post 9/11 World: A Report on Regional Discussions between the Science and Security Communities.* Washington, DC: National Academies Press.

NRC. 2008a. *Department of Homeland Security Bioterrorism Risk Assessment: A Call for Change.* Washington, DC: National Academies Press.

NRC. 2008b. *Emerging Cognitive Neuroscience and Related Technologies.* Washington, DC: National Academies Press.

NRC. 2008c. *Improving Democracy Assistance: Building Knowledge Through Evaluations and Research.* Washington, DC: National Academies Press.

NRC. 2009a. *The 2nd International Forum on Biosecurity: Summary of an International Meeting, Budapest, Hungary, March 30 to April 2, 2008.* Washington, DC: National Academies Press.

NRC. 2009b. *A Survey of Attitudes and Actions on Dual Use Research in the Life Sciences: A Collaborative Effort of the National Research Council and the American Association for the Advancement of Science.* Washington, DC: National Academies Press.

NRC. 2009c. *On Being a Scientist,* 3rd ed. Washington, DC: National Academies Press.

NRC/IOM. 2005. *Guidelines for Human Embryonic Stem Cell Research.* Washington, DC: National Academies Press.

NRC/IOM. 2007. *2007 Amendments to the National Academies' Guidelines for Human Embryonic Stem Cell Research.* Washington, DC: National Academies Press.

NRC/IOM. 2008. *2008 Amendments to the National Academies' Guidelines for Human Embryonic Stem Cell Research.* Washington, DC: National Academies Press.

NSABB (National Science Advisory Board for Biosecurity). 2007. *Proposed Framework for the Oversight of Dual Use Life Sciences Research: Strategies for Minimizing the Potential Misuse of Research Information.* Available at <http://oba.od.nih.gov/biosecurity/pdf/Framework%20f or%20transmittal%200807_Sept07.pdf>.

NSABB. 2008. *Strategic Plan for Outreach and Education on Dual Use Research Issues.* Available at <http://oba.od.nih.gov/biosecurity/PDF/FinalNSABBReportonOutreachandEducation-Dec102008.pdf>.

NSABB. 2009. *Enhancing Personnel Reliability among Individuals with Access to Select Agents.* Available at <http://oba.od.nih.gov/biosecurity/meetings/200905T/NSABB%20Final%20R eport%20on%20PR%205-29-09.pdf>.

NSC (National Security Council). 1969. *United States Policy on Chemical Warfare Program and Bacteriological/Biological Research Program.* National Security Decision Memorandum 35. Available at <http://www.state.gov/documents/organization/90919.pdf>.

OECD (Organisation for Economic Co-operation and Development). 2007. *OECD Best Practice Guidelines on Biosecurity for BRCs (Biological Resource Centers).* Paris: OECD. Available at <http://www.oecd.org/dataoecd/6/27/38778261.pdf>.

Ostfield, Marc L. 2009. Pathogen security: The illusion of security in foreign policy and biodefence. *International Journal of Risk Assessment and Management* 12(2-4):204-221.

Pesenti, Peter T. 2009. Presentation to National Academies Committee on Scientific Milestones for the Development of a Gene-Sequenced-Based Classification System for Oversight of Select Agents, September 3.

Rowe, Mary, Linda Wilcox, and Howard Gadlin. 2009. Dealing with—or reporting—"unacceptable" behavior (with additional thoughts about the "Bystander Effect"). *Journal of the International Ombudsman Association*, 2(1).

Sackett, Paul R., and Michael M. Harris. 1984. Honesty testing for personnel selection: A review and critique. *Personnel Psychology* 37:221-245.

Sackett, Paul R., Laura R. Burris, and Christine Callahan. 1989. Integrity testing for personnel selection: An update. *Personnel Psychology* 42:491-529.

Sackett, Paul R., and James E. Wanek. 1996. New developments in the use of measures of honesty, integrity, conscientiousness, dependability, trustworthiness, and reliability for personnel reliability. *Personnel Psychology* 49:787-829.

Sackett, Paul R., and Filip Lievens. 2008. Personnel selection. *Annual Review of Psychology* 59:419-450.

Schein, Edgar. 2001. Organizational culture and leadership. In *Classics of Organization Theory* (Jay Shafritz and J. Steven Ott, eds). Fort Worth, TX: Harcourt College Publishers.

Schön, D.A. 1973. *Beyond the Stable State: Public and Private Learning in a Changing Society.* Harmondsworth: Penguin.

Schulman, Paul R., Emory Roe, Michael van Eeten, and Mark de Bruijne. 2004. High reliability and the management of critical infrastructures. *Journal of Crisis and Contingency Management* 12(2):14-28.

Senge, Peter M. 1990. *The Fifth Discipline: The Art and Practice of the Learning Organization.* New York: Random House.

Sheppard, Blair H., Roy J. Lewicki, and John W. Minton. 1992. *Organizational Justice: The Search for Fairness in the Workplace.* New York: Lexington Books.

SIOP (Society for Industrial and Organizational Psychology). 2003. *Principles for the Validation and Use of Personnel Selection Procedures*, 4th ed. Bowling Green, OH: SIOP.

SIOP. 2009a. Types of Employment Tests. Available at <http://www.siop.org/Workplace/employ ment%20testing/testtypes.aspx>.

SIOP. 2009b. How Many U.S. Companies Use Employment Tests? Available at <http://www.siop. org/Workplace/employment%20testing/usingoftests.aspx>.

Stern, Jessica Eve. 2000. Larry Wayne Harris (1998). In *Toxic Terror: Assessing Terrorist Use of Chemical and Biological Weapons* (Jonathan B. Tucker, ed.), pp. 227-245. Cambridge, MA: MIT Press.

Sutton, Victoria. 2009. Survey finds biodefense anxiety—Over inadvertently violating regulations. *Biosecurity and Bioterrorism: Biodefense Strategy, Practice, and Science* 7(2):225-226.

Trans-Federal Task Force. 2009. *Report of the Trans-Federal Task Force on Optimizing Biosafety and Biocontainment Oversight.* July. Available at <http://www.hhs.gov/aspr/omsph/biosafetytask- force/biosafetycontainmentrpt092009.pdf>.

Tucker, Jonathan B., ed. 2000. *Toxic Terror: Assessing Terrorist Use of Chemical and Biological Weapons.* Cambridge, MA: MIT Press.

Turner, James T., and Michael G. Gelles. 2003. *Threat Assessment: A Risk Management Approach.* Binghamton, NY: The Haworth Press, Inc.

UN (United Nations). 2004. Security Council Resolution 1540 (S/RES/1540). Available at <http:// www.state.gov/t/isn/73519.htm>.

UN. 2008. Security Council Resolution 1810 (S/RES/1820). Available at <http://daccess-ods. un.org/access.nsf/Get?Open&DS=S/RES/1810%20(2008)&Lang=E&Area=UNDOC>.

USDA (U.S. Department of Agriculture). 2002. "USDA Security Policies and Procedures for Biosafety Level-3 Facilities." Departmental Manual 9610-001. Available at <http://www.ocio.usda.gov/directives/doc/DM9610-001.htm>.

USDA. 2005. "7 CFR Part 331 and 9 CFR Part 121: Agricultural Bioterrorism Protection Act of 2002; Possession, Use, and Transfer of Biological Agents and Toxins; Final Rule" (FR Doc. 05-5063). *Federal Register* 70(52, March 18), pp. 13242-13292.

Washington Post. 2009. "Cocaine Justice" (editorial). July 16, p. A18.

Weick, Karl E., and Kathleen M. Sutcliffe. 2001. *Managing the Unexpected: Assuring High Performance in an Age of Complexity.* San Francisco, CA: Jossey-Bass.

Weyant, Robbin S. 2009. E-mail communication to the committee. August 6.

Weyant, Robbin S., and Elizabeth Snyder. 2009. "National Academies Briefing: Security Risk Assessments for Possession, Use, and Transfer of Select Agents." Presentation to the Committee. June 29.

Wheelis, Mark, Lajos Rózsa, and Malcolm Dando, eds. 2006. *Deadly Cultures: Biological Weapons Since 1945.* Cambridge, MA: Harvard University Press.

White House. 2005. Revised Adjudicative Guidelines for Determining Eligibility for Access to Classified Information. Memorandum for William Leonard. December 29. Available at <http://www.fas.org/sgp/isoo/guidelines.html>.

White House. 2008. Reforming Processes Related to Suitability for Government Employment, Fitness for Contractor Employees, and Eligibility for Access to Classified National Security Information. Executive Order 13467 (June 30). Available at <http://www.fas.org/irp/offdocs/eo/eo-13467.htm>.

White House. 2009. "Strengthening Laboratory Biosecurity in the United States. Executive Order 13486 (January 9)" (FR Doc. E9-818). *Federal Register* 74(9, January 14), pp. 2289-2291.

WHO (World Health Organization). 2004. *Laboratory Biosafety Manual,* 3rd ed. WHO/CDS/CSR/LYO/2004.11. Geneva: WHO. Available at <http://www.who.int/entity/csr/resources/publications/biosafety/Biosafety7.pdf>.

WHO. 2006. *Biorisk Management: Laboratory Biosecurity Guidance.* WHO/CDS/EPR/2006.6. Geneva: WHO.

WHO. 2008. WHO recommendations concerning the distribution, handling, and synthesis of Variola virus DNA. Available at <http://www.who.int/csr/disease/smallpox/SummaryrecommendationsMay08.pdf>.

Wholey, Joseph, Harry P. Hatry, and Kathryn E. Newcomer, eds. 2004. *Handbook of Practical Program Evaluation,* 2nd ed. San Francisco, CA: Jossey-Bass.

Wilkie, Dana. 2004. "Select-Agent Security Clearance Stymies Research." *The Scientist* 18(10, 24 May):45.

WMD Commission (Commission on the Prevention of Weapons of Mass Destruction Proliferation and Terrorism). 2008. *World at Risk: The Report of the Commission on the Prevention of WMD Proliferation and Terrorism.* New York: Vintage Books.

World Bank. 2004. *Monitoring and Evaluation: Some Tools and Approaches.* Washington, DC: World Bank.

Appendixes

Appendix A

Committee Member and Staff Biographies

CHAIR

Rita R. Colwell, Ph.D., is Distinguished University Professor both at the University of Maryland at College Park and at Johns Hopkins University Bloomberg School of Public Health, president and chief executive officer of CosmosID, Inc., and senior advisor for Canon US Life Sciences, Inc. Her interests are focused on global infectious diseases, water, and health, and she is developing an international network to address emerging infectious diseases and water issues, including safe drinking water for both the developed and developing world. Dr. Colwell has held many advisory positions in the U.S. government, nonprofit science policy organizations, and private foundations, as well as in the international scientific research community. She is the recipient of 54 honorary doctorates including her alma mater, Purdue University. Dr. Colwell is a member of the National Academy of Sciences, the Royal Swedish Academy of Sciences, Stockholm, the American Academy of Arts and Sciences, and the American Philosophical Society. She was awarded the National Medal of Science by the President of the United States and the Order of the Rising Sun by the Emperor of Japan. Dr. Colwell holds a B.S. in bacteriology and an M.S. in genetics from Purdue University and a Ph.D. in oceanography from the University of Washington.

MEMBERS

Ronald M. Atlas, Ph.D., is professor of biology and public health and co-director of the Center for Health Hazards Preparedness at the University of Louisville. He earned his B.S. degree from the State University of New York at Stony Brook and his M.S. and Ph.D. degrees from Rutgers University. He was a postdoctoral fellow at the Jet Propulsion Laboratory where he worked on Mars Life Detection. He is a former president of the American Society for Microbiology (ASM) and currently is co-chair of ASM's Committee on Biodefense. He is also chair of the Wellcome Trust Strategy Committee on Infectious Disease and Population Health. Dr. Atlas is a former member of the Department of Homeland Security's Science and Technology Advisory Committee, NASA's Planetary Protection Board, the FBI Scientific Working Group on Bioforensics, and the NIH Recombinant DNA Advisory Committee. His early research focused on oil spills, and he discovered bioremediation as part of his doctoral studies. Later he turned to the molecular detection of pathogens in the environment, which forms the basis for biosensors to detect biothreat agents. He is author of nearly 300 manuscripts and 20 books. Dr. Atlas is a fellow of the American Academy of Microbiology and has received the ASM Award for Applied and Environmental Microbiology, the ASM Founders Award, the Edmund Youde Lectureship Award in Hong Kong, and an honorary D.Sc. from the University of Guelph.

John D. Clements, Ph.D., is a professor and chair of the Department of Microbiology and Immunology at Tulane University School of Medicine and director of the Tulane Center for Infectious Diseases. After receiving his doctorate in 1979 from the University of Texas Health Science Center at Dallas, Dr. Clements completed a National Research Council Associateship at Walter Reed Army Institute of Research in Washington, DC. In 1980, Dr. Clements was appointed as assistant professor in the Departments of Microbiology and Medicine at the University of Rochester School of Medicine in Rochester, NY. In 1982, Dr. Clements joined the faculty at Tulane University, being named as chair in 1999. From 2006 to 2009, he was vice dean for research in the School of Medicine. Dr. Clements maintains an active research program focused on development of vaccines against infectious diseases. Dr. Clements' research has been continuously funded from a variety of Public Health Service, Department of Defense, and pharmaceutical sources.

Joseph A. DiZinno, DDS, joined BAE Systems as a forensics expert after more than two decades of experience at the Federal Bureau of Investigation (FBI). Most recently, he was assistant director overseeing the FBI laboratory, where he led all laboratory cases and forensic responses. He also served as a special agent and participated in mitochondria DNA research, which led to its first application to forensic casework. Joe began his career with the FBI in 1986. He received a B.S. degree from the University of Notre Dame in 1975 and a D.D.S. degree from The Ohio State University in 1980.

Adolfo García-Sastre, Ph.D., is professor in the Department of Microbiology, Fischberg Chair and professor in the Department of Medicine, and co-director of the Global Health and Emerging Pathogens Institute at Mount Sinai School of Medicine in New York. He is also principal investigator for the Center for Research on Influenza Pathogenesis, one of six Centers of Excellence for Influenza Research and Surveillance funded by the National Institute of Allergy and Infectious Diseases. For the past 15 years, his research interest has been focused on the molecular biology of influenza viruses and several other negative strand RNA viruses. During his postdoctoral training in the early 1990s, he developed novel strategies for expression of foreign antigens by a negative strand RNA virus, influenza virus. His research has resulted in more than 200 scientific publications and reviews. He was among the first members of the Vaccine Study Section at the National Institutes of Health. In addition, Dr. García-Sastre is an editor for the *Journal of Experimental Medicine* and *PLoS Pathogens* and is a member of the editorial boards for the *Journal of Virology*, *Virology*, *Journal of General Virology*, and *Virus Research*. He has been a co-organizer of the international course on Viral Vectors (2001), held in Heidelberg, Germany, and sponsored by the Federation of European Biochemical Societies (FEBS), and of the first Research Conference on Orthomyxoviruses in 2001, held in Teixel, the Netherlands, and sponsored by the European Scientific Working Group on Influenza (ESWI).

Michael G. Gelles, Psy.D., is currently a senior manager with Deloitte Consulting LLP's federal practice in Washington, D.C., consulting in the areas of human capital management and systems and operations. Previously, he was the chief psychologist for the Naval Criminal Investigative Service (NCIS) for more than 16 years. He was the lead psychologist for the behavioral consultation team for the Criminal Investigations Task Force, and a member of numerous other task forces in the areas of workplace violence, insider threat, and ethics in consultation to national security. Prior to joining the NCIS in 1990, Dr. Gelles served as a clinical psychologist for the U.S. Navy. He is active in a number of professional organizations, including the American Psychological Association's Division of Police Psychology, the International Association of Chiefs of Police, the Psychology Services Section, the Society of Police and Criminal Psychology, and the Association of Threat Assessment Professionals. Dr. Gelles received his B.A. from the University of Delaware and his master's and doctorate degrees in psychology from Yeshiva University in New York. He completed his clinical and forensic training at the National Naval Medical Center and his advanced training at the Washington School of Psychiatry.

Robert J. Hawley, Ph.D., RBP, CBSP, serves as the senior biosafety professional for Midwest Research Institute's (MRI's) Mid-Atlantic Operations and is responsible for the technical oversight of all group biosafety, biosecurity and biosurety projects, and support staff. He performs incident investigations, biosafety threat and risk assessments, and threat and vulnerability and emergency requirements analyses at designated facilities to mitigate security and safety risks regarding the storage and handling of biological threat agents. Dr. Hawley also provides training in biological safety operations, maximum containment, recombinant DNA technology, and the science and safety of microbial agents and toxins for BSL-2, BSL-3, and BSL-4 operations. Before joining MRI in 2003, he worked at the U.S. Army Medical Research Institute of Infectious Disease (USAMRIID) for 15 years, where he was responsible for formulating, implementing, and interpreting USAMRIID's microbiological and industrial safety policies and procedures. Positions filled during his tenure at USAMRIID include safety and occupational health specialist, safety and occupational health manager, chief of the Safety and Radiation Protection Office, and Command Biological Safety Officer. He is a former president of the American Biological Safety Association (ABSA) and the Chesapeake Branch of ABSA.

Sally Katzen, J.D., is the executive managing director of The Podesta Group. From 2001 until 2008, she has been a visiting professor of law at George Washington University, University of Michigan, George Mason University, and the University of Pennsylvania. She has also taught American government courses at Smith College, Johns Hopkins University, and the University of Michigan (Washington Program). Before her teaching positions, she served as the administrator of the Office of Information and Regulatory Affairs in the Office of Management and Budget (OMB) (1993-1998), as the deputy director of the National Economic Council in the White House (1998-1999), and as the deputy director for management in OMB (1999-2001). Before her government service, she was a partner in the Washington, D.C., law firm of Wilmer, Cutler, and Pickering, specializing in administrative law and legislative matters. Ms. Katzen recently served on the NRC's Committee to Review the OMB Risk Assessment and the NRC's Committee on Evaluating Research Efficiency in the U.S. Environmental Protection Agency. She is a fellow of the National Academy of Public Administration. She earned her J.D. from the University of Michigan Law School.

Paul Langevin, P.Eng., is the Director of Laboratory Design for Merrick and Company and president of Merrick Canada ULC. He has more than 25 years of expertise in laboratory design, containment, and commissioning. Merrick is recognized around the world as one of the few architecture/engineering firms with significant expertise in the field of high containment. In addition to designing complete high-containment research facilities, Merrick offers the added benefit of design and fabrication of gloveboxes, environmental chambers, and remote material-handling systems that are often required to perform research in high containment conditions. Merrick has developed an extensive company portfolio of technically demanding projects for federal agencies and universities that demonstrate the full suite of corporate, team and individual skills, and abilities necessary for successful, high-quality design and full service engineering. The company's Facilities, Science & Technology (FaST) team provides engineering and architecture for specialized buildings, facilities, and equipment for the U.S. Department of Energy, the U.S. Department of Agriculture, the National Institute of Standards and Technology, the U.S. Air Force, the U.S. Army Corps of Engineers, and many other federal, university, R&D, and international clients who require specialized buildings and systems.

Todd R. LaPorte, Ph.D., is professor emeritus of political science and professor of the graduate school at the University of California, Berkeley. He teaches and publishes in the areas of organization theory, technology, and politics and on the organizational and decisionmaking dynamics of large, complex, technologically intensive organizations and the challenges of governance in a technological society. A principal of the Berkeley High Reliability Organization Project, his multi-disciplinary team studied the organizational aspects of safety-critical systems such as nuclear power, air traffic control, and nuclear aircraft carriers. His research concerns the evolution of large-scale organizations operating technologies demanding very high levels of operating reliable performance across a number of management generations and the relationship of large-scale technical systems and social complexity to political legitimacy. He has examined institutional challenges of multi-generation nuclear missions at Los Alamos National Laboratory, and, more recently, the U.S. National Polarorbiter Operational Environmental Satellite System. And, in parallel work, he has taken up questions of unconventional crisis management.

He was a fellow of the Woodrow Wilson International Center for Scholars, Smithsonian Institution, and elected to the National Academy of Public Administration. National Academy of Sciences service includes membership on the Board on Radioactive Waste Management, panels of the Committee on Human Factors and Transportation Research Board, and the Committees on Long Term Institutional Management of DOE Legacy Waste Sites: Phase Two and Principles and Operational Strategies for Staged Repository Systems. In addition to this committee, he currently serves on the NRC committee on the Assessment of Impediments to Interagency Cooperation on Space and Earth Science Missions. He served on the Secretary of Energy Advisory Board, Department of Energy, and he was on the Technical Review Committee, Nuclear Materials Technology Division, Los Alamos National Laboratory.

Stephen S. Morse, Ph.D., is professor of clinical epidemiology and was founding director of the Center for Public Health Preparedness at Columbia University. His professional interests include epidemiology of emerging infections (a concept he originated), international cooperation for infectious disease surveillance, and defense against bioterrorism. Dr. Morse returned to Columbia in 2000 after four years in government service as program manager at the Defense Advanced Research Projects Agency (DARPA), Department of Defense, where he co-directed the Pathogen Countermeasures Program and subsequently directed the Advanced Diagnostics Program. Before going to Columbia, he was assistant professor (virology) at The Rockefeller University in New York, where he remains an adjunct faculty member. Dr. Morse was chair and principal organizer of the 1989 NIAID/NIH Conference on Emerging Viruses (for which he originated the concept of emerging viruses/infections); served as a member of the Institute of Medicine (IOM)–National Academy of Sciences Committee on Emerging Microbial Threats to Health (and chaired its Task Force on Viruses), and was a contributor to its report, *Emerging Infections* (1992); was a member of the IOM's Committee on Xenograft Transplantation; currently serves on the Steering Committee of the IOM's Forum on Microbial Threats (formerly the Forum on Emerging Infections); and has served as an adviser to the World Health Organization and several government agencies, including the Centers for Disease Control and Prevention, Department of Defense, and the New York City Department of Health and Mental Hygiene. He is a fellow of the New York Academy of Sciences and a past chair of its microbiology section, a fellow of the American Academy of Microbiology, and an elected life member of the Council on Foreign Relations. He was the founding chair of ProMED (the nonprofit international Program to Monitor Emerging Diseases) and was one of the originators of ProMED-mail, an international network inaugurated by ProMED in 1994 for outbreak reporting and disease monitoring using the Internet.

Kathryn Newcomer, Ph.D., is a professor and director of the Trachtenberg School of Public Policy and Public Administration at the George Washington University where she is also the co-director of the Midge Smith Center for Evaluation Effectiveness, home of The Evaluator' Institute (TEI). She teaches public and nonprofit administration, program evaluation, research design, and applied statistics. She routinely conducts research and training for federal and local government agencies and nonprofit organizations on performance measurement and program evaluation, and she has designed and conducted evaluations for several U.S. federal agencies and dozens of nonprofit organizations.

Dr. Newcomer has published five books: *Improving Government Performance* (1989), *The Handbook of Practical Program Evaluation* (1994, 2nd edition 2004), *Meeting the Challenges of Performance-Oriented Government* (2002), *Getting Results: A Guide for Federal Leaders and Managers* (2005), and *Transformational Leadership: Leading Change in Public and Nonprofit Agencies* (2008), and a volume of New Directions for Public Program Evaluation, *Using Performance Measurement to Improve Public and Nonprofit Programs* (1997), and numerous articles in journals including *Public Administration Review*. She is a fellow of the National Academy of Public Administration and currently serves on the Comptroller General's Educators' Advisory Panel. She served as president of the National Association of Schools of Public Affairs and Administration (NASPAA) for 2006-2007. She has received two Fulbright awards, one for Taiwan (1993) and one for Egypt (2001-2004). She has lectured on performance measurement and public program evaluation in Ukraine, Brazil, Taiwan, and the United Kingdom.

Dr. Newcomer earned a B.S. in education and an M.A. in political science from the University of Kansas and her Ph.D. in political science from the University of Iowa.

Elizabeth Rindskopf Parker, J.D., has been dean of the McGeorge School of Law at the University of the Pacific since 2002. Her fields of expertise include national security and terrorism, international relations, public policy and trade, technology development and transfer, commerce, and civil rights and liberties litigation. During her tenure as dean at Pacific McGeorge, Dean Parker has begun several grant-supported initiatives involving national security and high school-to-professional educational pipeline programs, designed to support and encourage at-risk students in the preparation needed for success in college and law school.

Before becoming dean of Pacific McGeorge, she was the general counsel for the 26-campus University of Wisconsin System. Earlier she served as general counsel of the National Security Agency (1984-1989), principal deputy legal adviser at the U.S. Department of State (1989-1990), and general counsel for the Central Intelligence Agency (1990-1995). She has also been counsel to several major law firms, Bryan Care, LLP and Surrey & Morse, as well as serving as assistant director for mergers and acquisitions. She began her career as a Reginald Heber Smith Fellow at Emory University School of Law, and she later served as the director of the New Haven Legal Assistance Association, Inc. Later, while a cooperating attorney for the NAACP Legal Defense and Education Fund, she argued successfully twice before the Supreme Court of the United States.

She is a former chair and current member of the Advisory Board of the American Bar Association's Standing Committee on Law and National Security and a presidentially appointed member of the Public Interest Declassification Board, the Council on Foreign Relations, the Judicial Council of California's Access and Fairness Advisory Committee, and the Commission for Impartial Courts. She is also the chair of the Sacramento Chapter of the World Affairs Council and a member of the Board of the Sacramento Region Community Foundation.

Paul R. Sackett, Ph.D., is the Beverly and Richard Fink Distinguished Professor of Psychology and Liberal Arts at the University of Minnesota, Twin Cities. His research interests revolve around legal, psychometric, and policy aspects of psychological testing, assessment, and personnel decisionmaking in workplace and educational settings. He has served as the editor of *Personnel Psychology*, as president of the Society for Industrial and Organizational Psychology, as co-chair of the Joint Committee on the Standards for Educational and Psychological

Testing, as a member of the National Research Council's Board on Testing and Assessment, and as chair of the American Psychological Association's Board of Scientific Affairs. He served as chair of the Committee on the Youth Population and Military Recruitment from 1999-2003. He has a Ph.D. in industrial and organizational psychology from the Ohio State University.

NATIONAL RESEARCH COUNCIL STAFF

Adam P. Fagen, Ph.D., is a senior program officer with the Board on Life Sciences of the National Research Council. He came to the National Academies from Harvard University, where he most recently served as preceptor on molecular and cellular biology. He earned his Ph.D. in molecular biology and education from Harvard, working on issues related to undergraduate science courses; his research focused on mechanisms for assessing and enhancing introductory science courses in biology and physics to encourage student learning and conceptual understanding, including studies of active learning, classroom demonstrations, and student understanding of genetics vocabulary. Dr. Fagen also received an A.M. in molecular and cellular biology from Harvard, based on laboratory research in molecular evolutionary genetics, and a B.A. from Swarthmore College with a double-major in biology and mathematics. He served as co-director of the 2000 National Doctoral Program Survey, an on-line assessment of doctoral programs organized by the National Association of Graduate-Professional Students, supported by the Alfred P. Sloan Foundation, and completed by more than 32,000 students.

At the National Academies, Dr. Fagen has served as study director for *Bridges to Independence: Fostering the Independence of New Investigators in Biomedical Research* (2005), *Treating Infectious Diseases in a Microbial World: Report of Two Workshops on Novel Antimicrobial Therapeutics* (2006), *2007* and *2008 Amendments to the National Academies' Guidelines for Human Embryonic Stem Cell Research* (2007, 2008), *Understanding Interventions that Encourage Minorities to Pursue Research Careers: Summary of a Workshop* (2007), *Inspired by Biology: From Molecules to Materials to Machines* (2008), *Transforming Agricultural Education for a Changing World* (2009), and *Research at the Intersection of the Physical and Life Sciences* (2009). He is currently study director or responsible staff officer for several ongoing projects including the National Academies Summer Institute on Undergraduate Education in Biology and the National Academies Human Embryonic Stem Cell Research Advisory Committee.

Jo L. Husbands, Ph.D., is a scholar/senior project director with the Board on Life Sciences. Dr. Husbands managed the project that produced the 2004 report, *Biotechnology Research in an Age of Terrorism*, and directs the international activities following up on its recommendations, including the 2nd International Forum on Biosecurity held in Budapest in March 2008 and an international workshop on biosecurity education held in the fall of 2009. She represents the National Academy of Sciences on the Biosecurity Working Group of the InterAcademy Panel on International Issues, which also includes the academies of China, Cuba, the Netherlands (chair), Nigeria, and the United Kingdom. She managed a joint project with the American Association for the Advancement of Science (AAAS) that has carried out a survey of AAAS members in the life sciences to provide some of the first empirical data about scientists' knowledge of dual use issues and their attitudes toward their responsibilities to help mitigate the risks of misuse of scientific research.

From 2005 2008, Dr. Husbands was a senior project director with the Academies' Program on Development, Security, and Cooperation. From 1991-2005 she was the director of the Committee on International Security and Arms Control (CISAC) of the National Academy of Sciences and its Working Group on Biological Weapons Control. In 1998-1999 she also served as the first director of the Program on Development, Security, and Cooperation in the Academies' Office of International Affairs. From 1986-1991 she was director of the Academies' Project on Democratization and a senior research associate for its Committee on International Conflict and Cooperation. Before joining the National Academies, she worked for several Washington, D.C.-based nongovernmental organizations focused on international security.

Dr. Husbands is currently an adjunct professor in the Security Studies Program at Georgetown University. She is a member of the Advisory Council of Women in International Security, the International Institute for Strategic Studies, the Global Agenda Council on Illicit Trade of the World Economic Forum, and the editorial board of *International Studies Perspectives*. She is also a fellow of the International Union of Pure and Applied Chemistry. She holds a Ph.D. in political science from the University of Minnesota and a master's in international public policy (international economics) from the Johns Hopkins University School of Advanced International Studies.

Rita S. Guenther is a senior program associate with the Committee on International Security and Arms Control at the National Academies, where she has worked since September 2001. In her capacity as a senior program associate, Ms. Guenther has worked on several cooperative projects between U.S. and Russian scientists, including projects on the internationalization of the nuclear fuel cycle, indigenization of Russian nuclear material protection, control, and accounting programs, the future of the biosciences and biotechnology in Russia, and the Nuclear Cities Initiative. She was the project director for a joint National Academies–Russian Academies project on the *Future of the Nuclear Security Environment in 2015*, and she served as one of two National Academies' staff officers responsible for the completion of the unique, fast-track consensus study, *Strengthening U.S.–Russian Cooperation on Nuclear Nonproliferation*. In addition to her work on joint projects with the Russian

Academy of Sciences, she has also served on cooperative projects and activities with colleagues from India and Pakistan. Her experience also includes having served as a key staff member for the recent consensus report, *Improving Democracy Assistance: Building Knowledge through Evaluations and Research*, which provided findings and recommendations to the U.S. Agency for International Development's Democracy and Governance office. Rita speaks Russian and German, holds a master of arts in Russian studies from Georgetown University, and is currently a Ph.D. student of Russian history at Georgetown University. In 2007, she received a Fulbright-Hayes Fellowship to conduct archival research on her dissertation, which is provisionally titled, *Lived Liberalism: Local Expressions of Political Beliefs in Russia, 1860-1914*.

Carl-Gustav Anderson joined the Board on Life Sciences of the National Research Council in March 2009 and serves as senior program assistant. He received a B.A. in philosophy from American University in 2009, completing significant research projects on the status of empiricism in Tiantai Buddhism and the influence of modern science on the philosophy and development of the Kyoto School. He has focused his research interests on Southeast Asian interactions with Buddhism, with particular emphasis on the development of Buddhist philosophy of science and Buddhist approaches to feminism. He has worked closely with the All Women's Action Society (Malaysia), helping to engage young men in feminist dialogue and to present a feminist response to the unique identity politics of contemporary Malaysia.

Appendix B

Information on Briefings and Site Visits

As part of its information gathering, the committee held in-person briefings at two committee meetings and engaged in site visits at several institutions. Information about these meetings and site visits is contained below.

COMMITTEE MEETINGS

MEETING #1 AGENDA
MONDAY, JUNE 29, 2009
OPEN SESSION

11:00 a.m. **Representatives of project sponsor and interagency Working Group**

- H. Clifford Lane, *Director, Division of Clinical Research, National Institute of Allergy and Infectious Diseases, National Institutes of Health*

- Ben Petro, *Director for Biological Threat Reduction and Counterterrorism Policy, Office of the Coordinator for WMD Prevention, National Security Council, Executive Office of the President*

- Carol D. Linden, *Principal Deputy Director, Biomedical Advanced Research and Development Authority (BARDA), Office of the Assistant Secretary for Preparedness and Response, Department of Health and Human Services*

12:00 p.m. Overview of Select Agent Program *(lunch will be available)*

- Robbin Weyant, *Director, Division of Select Agents and Toxins, Coordinating Office for Terrorism Preparedness and Emergency Response, U.S. Centers for Disease Control and Prevention*

Respondents

- LouAnn C. Burnett, *Assistant Director and Biological Safety Officer, Vanderbilt Environmental Health & Safety, Vanderbilt University*

- Dennis W. Metzger, *Professor and Theopold Smith Alumni Chair; Director, Center for Immunology and Microbial Disease, Albany Medical College* [by teleconference]

1:45 p.m. Research and lessons from other sectors

- Kelley Krokos, *Senior Research Scientist, American Institutes for Research*

2:30 p.m. Break

2:45 p.m. Laboratory security

- **Overview of biosafety, biosurety, and biosecurity:** Robert J. Hawley (committee member), *Senior Advisor, Science, Mid-Atlantic Operations, Midwest Research Institute*

- **Implementation of biosurety:** Jeffrey Adamovicz, *Principal Science Advisor, Center for Biological Safety and Security, Mid-Atlantic Operations, Midwest Research Institute*

- **Physical and operational security solutions:** Paul Langevin (committee member), *Director of Laboratory Design, Merrick and Company*

3:45 p.m.   **Personnel reliability**

- **Lessons from polygraph testing:** Kevin R. Murphy, *Professor of Psychology, Penn State University; Member, Committee to Review the Scientific Evidence on the Polygraph, National Research Council*

- **Establishing a culture of trust:** Mary P. Rowe, *Ombudsperson, Massachusetts Institute of Technology; Adjunct Professor of Negotiation and Conflict Management, MIT Sloan School of Management* [by videoconference]; Linda Wilcox, *Ombudsperson, Harvard Medical School, Harvard Dental School, and Harvard School of Public Health* [by videoconference]

4:45 p.m.   **General Discussion and Public Comment**

5:30 p.m.   End of open session

TUESDAY, JUNE 30, 2009
OPEN SESSION

9:00 a.m.   **Lessons from the nuclear community:** Richard A. Meserve, *President, Carnegie Institution for Science; Former Chairman, Nuclear Regulatory Commission*

9:45 a.m.   **Lessons from workplace violence**

- Robert A. Fein, *Department of Psychiatry, McLean Hospital and Harvard Medical School*

- Bryan Vossekuil, *National Violence Prevention and Study Center; Former Executive Director, National Threat Assessment Center, U.S. Secret Service (retired)*

10:30 a.m.   End of open session

MEETING #2 AGENDA

MONDAY, AUGUST 10, 2009
OPEN SESSION

8:30 a.m.   **Introduction and overview of open session** *(Breakfast will be available)*

- Rita R. Colwell (Committee Chair), *Distinguished University Professor, University of Maryland, College Park and Johns Hopkins University Bloomberg School of Public Health; President and CEO, CosmosID, Inc.*

8:45 a.m.   **Briefings on relevant issues from the aviation industry**

- Diane L. Damos, *President, Damos Aviation Services, Inc.*

- Bruce Landry, *Safety and Certification Specialist, Airport Safety and Operations Division, Federal Aviation Administration*

10:00 a.m. **Briefings on Security Risk Assessments in the broader security clearance context**

- M. Colleen Crowley, *Executive Program Director, Policy, Research, and Agency Support Program, Federal Investigative Services Division, U.S. Office of Personnel Management*

- J. William Leonard, *Principal, The Leonard Consulting Group, LLC*

- Sheldon I. Cohen, *Sheldon I. Cohen & Assoc., Attorneys At Law*

11:15 a.m. **General discussion** *(Lunch will be available)*

12:30 p.m. Van transportation to site visit locations

1:30 p.m.   **Site visits to select agent laboratories**

- **George Mason University:** National Center for Biodefense and Infectious Diseases, Manassas, VA

- **U.S. Department of Agriculture:** National Plant Germplasm and Biotechnology Laboratory, Beltsville, MD

5:30 p.m.   End of open session; van transportation back to Keck Center

TUESDAY, AUGUST 11, 2009
CLOSED SESSION—COMMITTEE MEMBERS AND STAFF ONLY

WEDNESDAY, AUGUST 12, 2009
OPEN SESSION

9:00 a.m.   **Discussion of report from Executive Order Working Group**
(Breakfast available)
- Carol D. Linden, *Principal Deputy Director, Biomedical Advanced Research and Development Authority (BARDA), Office of the Assistant Secretary for Preparedness and Response, Department of Health and Human Services*

10:00 a.m.   End of open session

SITE VISITS

Members of the committee and staff conducted site visits and on-site conversations with individuals affiliated with select agent laboratories and other secure facilities at several institutions. Participants at the site visit locations are listed.

1. New England Regional Center of Excellence (NERCE) for Biodefense and Emerging Infectious Diseases Research, Harvard Medical School, Boston, MA; visited August 4, 2009
   - Gerald Beltz, *Associate Director for Research, NERCE*
   - Christine Anderson, *Assistant Director, BSL-3 Animal and Tissue Culture Core Laboratory, NERCE*
   - Mary Corrigan, *Associate Director, Environmental Health and Safety, Harvard University*
   - Robert A. Dickson, *Associate Director of Operations, Harvard Medical School*
   - Sarah Heninger, *Assistant Director, Microbiology & Animal Resources Core Laboratory, NERCE*
   - Jeff M. Sco, *Director of Research Compliance, Harvard Medical School*

2. MIT Nuclear Reactor Laboratory (NRL), Massachusetts Institute of Technology (MIT), Cambridge, MA; visited August 4, 2009
   - John Bernard, *Director of Reactor Operations, NRL*
   - David Carpenter, *Doctoral Student in Nuclear Science and Engineering, MIT*

- Patricia Drooff, *Officer, Reactor-Radiation Protection Program, Environment, Health and Safety Office, MIT*
- Edward S. Lau, *Superintendent, NRL*
- William B. McCarthy, *Deputy Director, Reaction Radiation Protection Program, Environment, Health and Safety Office, MIT*
- Thomas H. Newton, Jr., *Associate Director for Engineering, NRL*
- Kathleen A. O'Connell, *Senior Administrative Assistant, NRL*

3. MIT Environment, Health and Safety Office, Massachusetts Institute of Technology, Cambridge, MA; visited August 4, 2009
   - Claudia A. Mickelson, *Deputy Director, Biosafety Program*

4. National Plant Germplasm and Biotechnology Laboratory, Center for Plant Health, Science, and Technology, U.S. Department of Agriculture, Beltsville, MD; visited August 10, 2009
   - Laurene Levy, *Laboratory Director*
   - Wayne P. Claus, *Facility Manager*
   - Renee DeVries, *Quality Manager*
   - Joseph P. Kozlovac, *Agency Biosafety Officer, Animal Production & Protection, Agricultural Research Service*

5. National Center for Biodefense and Infectious Diseases (NCBID), George Mason University (GMU), Manassas, VA; visited August 10, 2009
   - Charles Bailey, *Distinguished Professor of Biology; Laboratory Director, Biomedical Research Laboratory; Executive Director, NCBID*
   - Saira Ahmad, *Doctoral Student*
   - Lilian S. Amer, *Masters Student*
   - John H. Blacksten, *Director, GMU Office of Media & Public Relations*
   - Calvin Carpenter, *Deputy Director and Chief of Contract Services, Biomedical Research Laboratory, NCBID*
   - Jessica H. Chertow, *Doctoral Student*
   - Myung-Chul Chung, *Research Associate Professor, NCBID*
   - Meghan W. Durham-Colleran, *Doctoral Student*
   - Suhua Han, *Laboratory and Research Specialist, NCBID*
   - Jessica Kidd, *Laboratory and Research Specialist, NCBID*
   - Nathan Manes, *Postdoctoral Research, NCBID*
   - Beth McKenney, *Masters Student*
   - Marjorie Z. Musick, *Manager, GMU Office of Media & Public Relations*
   - Tony Pierson, *Doctoral Student*
   - Kathleen Powell, *Administrative Specialist, NCBID*

- Meena Rajan, *HR Operations Coordinator, GMU Human Resources & Payroll*
- Ian Reynolds, *Human Resources Consultant, GMU Human Resources & Payroll*
- Diann Stedman, *Director, Laboratory Safety, GMU Environmental Health & Safety Office*
- Anne Taylor, *Technical Operations Manager, NCBID*
- Patty Theimer, *Senior Fiscal Technician, GMU Life Sciences*
- Monique L. van Hoek, *Assistant Professor of Molecular and Microbiology*
- Anne B. Verhoeven, *Doctoral Student*
- Paul R. Wieber, *Security Manager, Prince William Campus, GMU Police*
- James D. Willett, *Professor of Molecular and Microbiology*
- Ronald Witt, *Director of Maintenance and Operations, Biomedical Research Laboratory, NCBID*

Appendix C

Abbreviations and Acronyms

AAAS	American Association for the Advancement of Science
AAMC	Association of American Medical Colleges
ABSA	American Biological Safety Association
ADA	Americans with Disabilities Act
AG	The Australia Group
AMA	American Management Association
ANACI	Access National Agency Check and Inquiries
APHIS	Animal and Plant Health Inspection Service
ARO	Alternate Responsible Official
ARS	Agricultural Research Service
ASTM	formerly American Society for Testing and Materials
BMBL	Biological Safety in Microbiological and Biomedical Laboratories
BRL	Biomedical Research Laboratory (at George Mason University)
BSAT	biological select agents and toxins
BSATAC	Biological Select Agents and Toxins Advisory Committee
BSC	biosafety cabinet
BSL	biosafety level
BTRA	bioterrorism risk assessment
BWC	Biological and Toxin Weapons Convention
CDC	Centers for Disease Control and Prevention
CEN	European Committee for Standardization / Comité Européen de Normalisation

163

CFR	Code of Federal Regulations
CIRP	Critical Incident Response Program (for the Airline Pilots Association)
CJIS	Criminal Justice Information Services
CVB	Center for Veterinary Biologics
DHB	Defense Health Board
DHS	Department of Homeland Security
DNA	deoxyribonucleic acid
DOD	Department of Defense
DOE	Department of Energy
DOJ	Department of Justice
DSB	Defense Science Board
EAP	employee assistance program
EO	Executive Order
ESCRO	Embryonic Stem Cell Research Oversight (Committee)
FAQ	frequently asked questions
FASEB	Federation of American Societies for Experimental Biology
FBI	Federal Bureau of Investigation
GAO	Government Accountability Office
GMO	genetically modified organism
GMU	George Mason University
HHS	Department of Health and Human Services
IACUC	Institutional Animal Care and Use Committee
IAEA	International Atomic Energy Agency
IBC	Institutional Biosafety Committee
III	Interstate Identification Index
IRB	Institutional Review Board
ISATTAC	Intragovermental Select Agents and Toxins Technical Advisory Committee
LRN	Laboratory Response Network
MCMI	Millon Clinical Multiaxial Inventory
MMPI	Minnesota Multiphasic Personality Inventory
MUF	materials unaccounted for
NAC	National Agency Check

NACI	National Agency Check and Inquiries
NACIC	National Agency Check and Inquiries and Credit
NACLC	National Agency Check with Local Agency Check and Credit Check
NAE	National Academy of Engineering
NBACC	National Biodefense Analysis and Countermeasures Center
NBL	National Biocontainment Laboratory
NCBID	National Center for Biodefense and Infectious Diseases (at George Mason University)
NELP	National Employment Law Project
NIAID	National Institute of Allergy and Infectious Diseases
NIH	National Institutes of Health
NIOSH	National Institute for Occupational Safety and Health
NRC	National Research Council
NSABB	National Science Advisory Board for Biosecurity
NSF	National Science Foundation
OECD	Organization for Economic Cooperation and Development
OIG	Office of Inspector General
OPM	Office of Personnel Management
OSTP	Office of Science and Technology Policy
PAI	Personality Assessment Inventory
PI	principal investigator
PPE	personal protective equipment
PRP	personnel reliability program
PSC	Professional Standards Committee (for the Airline Pilots Association)
RAC	Recombinant DNA Advisory Committee
RBL	Regional Biocontainment Laboratory
RCE	Regional Center of Excellence (for Biodefense and Emerging Infectious Diseases)
RCR	responsible conduct of research
RO	Responsible Official
SACHRP	Secretary's Advisory Committee on Human Research Protections
SIOP	Society for Industrial and Organizational Psychology
SRA	Security Risk Assessment
SSBI	Single Scope Background Investigation
TSA	Transportation Security Administration

UFAS   Uniform Federal Accessibility Standards
UFC   Unified Facilities Criteria
UK   United Kingdom
UL   Underwriters Laboratories
UN   United Nations
UNOLS   University-National Oceanographic Laboratory System
UNSCR   United Nations Security Council Regulation
USAMRIID   U.S. Army Medical Research Institute of Infectious Diseases
USA PATRIOT   Uniting and Strengthening America by Providing
                Appropriate Tools Required to Intercept and Obstruct
                Terrorism
USC   United States Code
USDA   U.S. Department of Agriculture
UTMB   University of Texas Medical Branch at Galveston

WHO   World Health Organization
WMD   weapons of mass destruction

Appendix D

Application for Security Risk Assessment

FBI Form FD-961
Bioterrorism Preparedness Act: Entity / Individual Application

FD-961 (Rev. 08-31-06)   OMB No. 1110-0039-Exp 08-31-09

[Reset Form]

FEDERAL BUREAU OF INVESTIGATION
BIOTERRORISM PREPAREDNESS ACT: ENTITY / INDIVIDUAL INFORMATION

Section I: Entity Information (Identical to that indicated on the CDC or APHIS registration application)
1. Legal Name of Entity:

2. Address: (**Not** a post office box) Street City   County   State   Zip Code

3. Type of Entity:

☐ Public   ☐ Government

☐ Other (i.e. Non-Profit, Private Academic, and Commercial)
*** **Indicate if you are a** ☐ **corporate officer,** ☐ **board of director, and/or** ☐ **stock holder.**

Section II: Individual Information

4. Full Name (Last, First, Middle)	5. Date of Birth (Month, Day, Year)	6. Social Security Number
4a. Aliases/Maiden Name:		

7. Residence Address: (No., Street, City, State, Zip Code)	8. Sex: ☐ Male ☐ Female

9. Place of Birth (City, State or Foreign Country)

*If not born in the United States please complete questions on page 2 titled Foreign Born Information.

10. Race:

☐ White

☐ Black or African   ☐ Hispanic or Latino

☐ Asian/ Native Hawaiian   ☐ American Indian or
   Alaska Native   Pacific Islander

11. Unique Identifier Number (Supplied by APHIS or CDC):

12. Certifications (All questions must be answered "Yes" or "No" in the box provided)

*Title 18 Section 1001 of the U.S. Code provides that knowingly falsifying or concealing a material fact is a felony that may result in fines or imprisonment for not more than 5 years or both.

12a. Are you under indictment or information in any court for a felony, or any crime, for which the judge could imprison you for more than one year? ☐ Yes ☐ No	12b. Have you been convicted in any court for a crime, for which the judge could have imprisoned you for more than one year, even if you received a shorter sentence including probation? ☐ Yes ☐ No
12c. Are you a fugitive from justice? ☐ Yes ☐ No	12d. Are you an unlawful user of any controlled substance (as defined in Section 102 of the Controlled Substance Act [21 U.S.C. 802])? ☐ Yes ☐ No
12e. Have you ever been adjudicated as a mental defective or been committed to any mental institution? If yes, a complete copy of medical records regarding the commitment will be required. ☐ Yes ☐ No	12f. Are you an alien illegally or unlawfully in the United States? ☐ Yes ☐ No
12g. Are you an alien who has been lawfully admitted for permanent residence or a naturalized citizen? If yes, please complete page 2 of the application. ☐ Yes ☐ No	12h. Have you been discharged from the Armed Services of the United States under dishonorable conditions? ☐ Yes ☐ No

I certify that the above answers are true, correct and complete. I understand that the making of a false oral or written statement is a crime. Signature	Date: 12/10/2009

1   [Save Form]   [Print]   [Next Page]

Previous Page

Foreign Born Information

This page must be completed by any individual answering YES to question 12g of page 1. All questions **MUST** be answered. Be sure to include all alien or admission numbers for question 9.

13.   Country of Citizenship:

14.   Mother's Full Name:

15.   Father's   Full Name:

16.   Date of Entry to the United States:

17.   Place of Entry:

18.   Immigration Status at Entry:

19.   Current Immigration Status:

20.   Date Status Expires, if Applicable:

21.   Alien Number or Admission Number (9-11 digits):

Alien registration numbers are issued by the Bureau of Immigration and Customs Enforcement for individuals who are granted permanent legal resident or a naturalized citizen status in the U.S. Other situations that individuals would have an alien registration number include the following: Employment Authorization cards, Temporary Resident cards, Border Crossing cards, I-94 or Visa numbers. If this number is not available please provide an explanation. If born to US citizen serving a military or diplomatic post in a foreign country please provide a copy of the US born abroad birth certificate.

Save Form   Print   Next Page

Previous Page

Section III: Consent

By signing this form, I hereby authorize the U.S. Department of Justice to obtain any information relevant to assessing my suitability to access, possess, use, receive or transfer select agents and toxins from any relevant source, including, but not limited to, individuals, public sources, and government sources. This information may include, but is not limited to, biographical, financial, law enforcement and intelligence information.

I further authorize any individuals having information pertinent to such an assessment to release such information to a duly accredited representative of the U.S. Department of Justice. The authorization set forth in this paragraph is valid for five (5) years from the date on which this form is signed.

I further authorize the U.S. Department of Justice to disclose any records, results or information relating to, or obtained in connection with, my security risk assessment to: the U.S. Department of Agriculture; the Department of Health and Human Services; any agency contractors assisting in the determination of risk; and responsible officers or other appropriate personnel of pertinent entities.

I further authorize the release of records, results or information relating to, or obtained in connection with my security risk assessment to any law enforcement or intelligence authority or other federal, state or local entity with relevant jurisdiction where such information reveals a risk to human, animal and/or plant health or national security.

I further authorize disclosure of records results or information relating to, or obtained in connection with my security risk assessment to organizations or individuals, both public and private, if deemed necessary, in the sole discretion of the U.S. Department of Justice, to elicit information or cooperation from the recipient for use in assessing my suitability to access, possess, use, receive or transfer select agents and toxins.

I further authorize release of records, results or information relating to, or obtained in connection with my security risk assessment to laboratories, universities, individuals, or other entities, both public and private, responsible for making security assessments, employment and/or licensing determinations and suitability or security decisions when the information is relevant to an assessment of my suitability to access, possess, receive, use, or transfer agents or toxins

I understand that this is a legally binding document and false statements provided by me are violations of federal law and may lead to criminal prosecution or other legal action.

_____   12/10/2009
PRINTED NAME   DATE

_____   Save Form   Print
SIGNATURE

3